Arms and the State

Modern Economic and Social History Series

General Editor: Derek H. Aldcroft

Titles in this series include:

Studies in the Interwar European Economy
Derek H. Aldcroft

Whatever Happened to Monetarism?
Economic Policy Making and Social Learning in the United Kingdom Since 1979
Michael J. Oliver

Disillusionment or New Opportunities?
The Changing Nature of Work in Offices, Glasgow 1880–1914
R. Guerriero Wilson

Raleigh and the British Bicycle Industry
An Economic and Business History, 1870–1960
Roger Lloyd-Jones and M.J. Lewis

Battles for the Standard
Bimetallism and the Spread of the Gold Standard in the Nineteenth Century
Ted Wilson

The British Footwear Industry
Peter R. Mounfield

Trade Unions and the Economy: 1870–2000
Derek H. Aldcroft and Michael J. Oliver

The British Electronics Industry
Technical Promise and Market Failure, 1930–2000
John F. Wilson

The Seaside, Health and the Environment in England and Wales since 1800
John Hassan

Exchange Rates and Economic Policy in the 20th Century
Edited by Ross E. Catterall and Derek H. Aldcroft

Arms and the State

Sir William Armstrong and the Remaking of British Naval Power, 1854–1914

MARSHALL J. BASTABLE

Routledge
Taylor & Francis Group
LONDON AND NEW YORK

First published 2004 by Ashgate Publishing

Published 2016 by Routledge
2 Park Square, Milton Park, Abingdon, Oxon OX14 4RN
605 Third Avenue, New York, NY 10017

Routledge is an imprint of the Taylor & Francis Group, an informa business

British Library Cataloguing in Publication Data
Bastable, Marshall J.
Arms and the state : Sir William Armstrong and the remaking of British naval power, 1854–1914. – (Modern economic and social history)
1. Armstrong of Cragside, William George Armstrong, Baron, 1810–1900 2. Sir W.G. Armstrong Whitworth & Co. – History
3. Weapons industry – Great Britain – History – 19th century
4. Great Britain – History, Naval – 19th century
I. Title
338.4'76234'0941'09034

Library of Congress Cataloging-in-Publication Data
Bastable, Marshall J., 1943–
Arms and the state : Sir William Armstrong and the remaking of British naval power, 1854–1914 / Marshall J. Bastable.
p. cm. – (Modern economic and social history)
Includes bibliographical references and index.
ISBN 0-7546-3404-3 (alk. paper)
1. Defense industries–Great Britain–History. 2. Armstrong, W.G. (William George), Baron, 1810–1900. 3. Great Britain. Royal Navy–Ordnance and ordnance stores–History. 4. Great Britain. Royal Navy–History. I. Title. II. Modern economic and social history series

HD9743.G72B37 2003
338.7'623418'094109034–dc21

2003042161

ISBN 13: 978-0-7546-3404-1 (hbk)
ISBN 13: 978-1-138-25171-7 (pbk)

Typeset by Express Typesetters Ltd, Farnham

To RONNIE

and for

JO (1981–2003)

Contents

Modern Economic and Social History Series
General Editor's preface

Economic and social history has been a flourishing subject of scholarly study during recent decades. Not only has the volume of literature increased enormously but the range of interest in time, space and subject matter has broadened considerably so that today there are many sub-branches of the subject which have developed considerable status in their own right.

One of the aims of this new series is to encourage the publication of scholarly monographs on any aspect of modern economic and social history. The geographical coverage is world-wide and contributions on non-British themes will be especially welcome. While emphasis will be placed on works embodying original research, it is also intended that the series should provide the opportunity to publish studies of a more general and thematic nature which offer a reappraisal or critical analysis of major issues of debate.

Derek H. Aldcroft

University of Leicester

Abbreviations

The name of the company changed as it grew: from 1859 to 1882 it was the Sir William G. Armstrong and Company; from 1882 until 1896 it was the Sir William G. Armstrong, Mitchell & Company; and thereafter it was called the Sir W.G. Armstrong, Whitworth & Company. However, the company was well-known by its unofficial name, Elswick, after the area of Newcastle where its factories, shipyards and head office were located. For convenience, and readily to distinguish Armstrong the man from his company, Elswick is used most often in this study to refer to the company.

Adm.	Admiralty Papers, Public Record Office (PRO)
D.B.B.	Dictionary of Business Biography
D.N.B.	Dictionary of National Biography
F.O.	Foreign Office Papers, PRO
N.L.W.	National Library of Wales
N.M.M.	National Maritime Museum, London
S.O.A.S.	School of Oriental and African Studies, London
T&W	Tyne & Wear County Archive, Newcastle
W.O.	War Office Papers, PRO

Photograph credits

All photographs appear with the permission of the copyright holders.

This book is about the invention, development and proliferation of the first weapons of mass destruction. We should keep in mind that making great armaments is less than half the story.

> The horse glides, menacing, advancing toward the centre of the city ... the arms clashed loud inside its belly. Nevertheless, heedless, blinded by frenzy, we press right on and set the inauspicious monster inside the sacred fortress. Even then can Cassandra chant of what will come with lips the gods had doomed to disbelief.
>
> (Virgil, *The Aeneid*)[1]

> An uncertain red glow spreads along the skyline from one end to the other. It is in perpetual movement, punctuated with the bursts of flame from the nozzles of the batteries. Balls of light rise up high above it, silver and red spheres which explode and rain down in showers of red, white, and green stars. French rockets go up, which unfold a silk parachute to the air and drift slowly down. They light up everything as bright as day, their light shines on us and we see our shadows sharply outlined on the ground. They hover for the space of a minute before they burn out. Immediately fresh ones shoot up in the sky, and again green, red, and blue stars.
> 'Bombardment,' says Kat.
>
> Erich Maria Remarque, *All Quiet on the Western Front*[2]

> The artillery barrage is deafening. When the air is still, the din can be heard faintly in London and Paris. Sometimes the pounding lasts for days. In June 1916 at the Somme it continues for seven days and nights. Field artillery, medium artillery, and heavy howitzers. The fifteen-inch-caliber gun of the British can fire a shell of fourteen hundred pounds. 'Big Bertha' of the Germans, with a caliber of seventeen inches, can project a missile weighing over a ton. At Verdun in 1916 the Germans bring in thirteen of those twenty-ton monsters. Each is moved into position by nine tractors; a crane is required to insert the shell. The impact of this shell annihilates buildings; it shatters windows in a two-mile radius. In August 1914 these huge machines of war had demolished the purportedly impregnable forts of Liège. As the Krupp guns 'walked' their shells towards the final target, Belgian defenders inside the forts went mad.
>
> Modris Eksteins, 'Rites of War: Battle Ballet'[3]

> The earth rocked and the sky boiled like a gigantic cauldron. ... Head and ears ached violently, and we could only make ourselves understood by shouting a word at a time. ... In this neighbourhood of villages, meadows, woods, and fields there was literally not a bush or a tiniest blade of grass to be seen. ... And yet the strangest thing of all was not the horror of the landscape in itself, but the fact that these scenes, such as the world had never known before, were fashioned by men who intended them to be a decisive end to war. ... And it seemed that man, on this landscape he had himself created, became different, more mysterious and hardy and callous.
>
> (Ernst Jünger, *Storm of Steel*)[4]

> The moral is obvious: it is that great armaments lead inevitably to war. ... The enormous growth of armaments in Europe, the sense of insecurity and fear caused by them – it was these that made war inevitable.
>
> (Sir Edward Grey, *Twenty-Five Years*)[5]

> Lomax: Well, the more destructive war becomes, the sooner it will be abolished, eh?
> Undershaft: Not at all. The more destructive war becomes the more fascinating we find it.
>
> (Bernard Shaw, *Major Barbara*)[6]

Notes

1. *The Aeneid of Virgil*, trans. Allen Mandelbaum, Berkeley: University of California Press, 1971, p. 37.
2. Erich Maria Remarque, *All Quiet on the Western Front*, New York: Fawcett Crest, 1975, pp. 58–9.
3. Modris Eksteins, 'Rites of War: Battle Ballet', *The Rites of Spring: The Great War and the Birth of the Modern Age*, Toronto: Lester & Orpen Dennys, 1989, pp. 139–40.
4. Ernst Jünger, *Storm of Steel: From the Diary of a German Storm-Troop Officer on the Western Front*, trans. Basil Creighton, London: Chatto and Windus, 1929, *passim.*
5. Sir Edward Grey, *Twenty-Five Years*, vol. 1, New York: Frederick Stokes, 1925, pp. 89–90.
6. Bernard Shaw, *Major Barbara*, Harmondsworth, Middlesex: Penguin Books, 1960, Act I, p. 70.

Introduction: theories, models and approaches

The manufacture and sale of armaments became big business in the late nineteenth century. Driven by technological innovation and the commercialization of armaments production for a global market, armaments companies were at the forefront of the general scramble of industrialists to amalgamate and integrate their businesses. The political, economic and cultural impact of the new armaments industry between 1890 and 1914 is difficult to exaggerate. Naval armaments in particular increased their economic importance, moving up domestic and international political agendas, stirring imperialist dreams and confounding strategists. The Anglo-German naval arms race of 1905–14 was just the greatest of several such races throughout the world prompted by nationalist ambitions and imperialist dreams. By 1914, armaments had become a major industrial and military force in the world, and the armaments business was in the hands of a few giant firms. One of those firms was the British firm of Sir William Armstrong, Whitworth and Company.

The primary purpose of this study is to reconstruct the history of that company from the archival and printed primary sources, from its origins in the Crimean War to its high point at the outbreak of the First World War, and to set that story in its technological, political and international contexts.

Much has been written about the company for the period 1880–1914, when information about the company is more readily available. This study includes new research on that period, but it also brings to light far more about the earlier period of the company's development which highlights the technological, political and international nature of the story.

The research historian must first track down the basic facts about what happened in precise chronological order, who were the significant individuals, organisations and institutions involved, why those events happened, and what the impact was of those actions. The reconstructed stories provide the substance for the more complex task of linking particular episodes to the significant broader political, economic and cultural contexts, but they ought also to be granted individual recognition. Yet to cast aside all models or theories would leave history a random string of unrelated events and cut off the past entirely. On the other hand, to impose a single, all-encompassing model or perspective on such a large enterprise would rob its history of its rich complexity. Theories and models should not be allowed to exclude the full story.

Accordingly, the aim here is to tell the story of armaments technology and the armaments business, referring from time to time to certain theories and models of technological innovation, British entrepreneurship, the relationship between armaments engineers and businessmen and the British state and its foreign policies. That armaments have an impact upon domestic politics and culture on the one hand, and the balance of global power and international relations on the other, is obvious. Less salient but nevertheless real is the way political, cultural and global changes influence technological innovations from the general decision to develop armaments to solve specific problems, down to the specific designs. Put this way, we see that technology is designed to satisfy specific human desires within changing historical contexts. There is thus a delicate balance between the decisions of engineers, inventors, businessmen and politicians, and the global structure in which they act. The following highlights of the Armstrong story will serve as a prerequisite for an introductory discussion of various theories and models pertinent to this study.[1]

The basic story

Sir William Armstrong founded his armaments company in 1859 to manufacture and sell his new artillery under an exclusive contract with the British government. His influence in the company remained dominant until his death in 1900, after which Sir Andrew Noble, a founding member of the firm, took command until his death in 1915. Armstrong was a typical Victorian entrepreneur: a brilliantly inventive, self-taught engineer who built up a large and successful business from his own technological inventions. Born in Newcastle in 1810, Armstrong's formative years fell in the middle of the great Industrial Revolution, when British entrepreneurs were transforming the economic and social landscape at home and abroad with their technological inventions and business ventures. As a young man, Armstrong was sent off to London to study law, but his passion was engineering and, amidst the great railway frenzy of the 1840s, he plunged into the study and the business of mechanical engineering, and established a successful engineering company in Newcastle which manufactured hydraulic cranes of Armstrong's design.

The Crimean War directed Armstrong's attention to armaments, and he quickly designed, tested and developed Britain's first practical breech-loaded, rifled artillery piece. The gun proved itself against competitors and Armstrong found himself in the armaments business in 1859, when Napoleon III launched his war to drive Austria out of northern Italy, rekindling British fears of a French invasion. The War Office and Admiralty bought as many guns as Armstrong could make, and the state arsenal at Woolwich was given over to Armstrong to develop larger versions of his design. The fact that the Royal Navy put Armstrong guns on its warships established Armstrong's international reputation as a pre-eminent gun maker. When the crisis with France passed, the British government cancelled the contract, in 1863, and Armstrong turned to the global markets.

Armstrong's Newcastle partners convinced him to found the Elswick Ordnance Works in 1863 in order to sell his new guns abroad, and indeed governments from Europe and overseas clamoured to buy the famous Armstrong guns. Armstrong expanded into building small gunboats but his main product was his monster guns for battleships. The business grew rapidly throughout the 1860s and 1870s. His quick-firing guns on small but fast warships proved very successful in the 1880s as a cheap counter to the great lumbering battleships. With the return of the battleship in the 1890s, Armstrong's gun mounts proved crucial to manipulate the large naval guns. The business expanded through mergers and acquisitions throughout the 1880s and 1890s, reaching a climax with the purchase of Sir Joseph Whitworth's steel company in 1897. At the time of Armstrong's death in 1900, the Sir William G. Armstrong, Whitworth Company was one of Britain's most important industrial enterprises and one of the world's largest manufacturers of armaments. Armstrong's business success was based on technological innovation and an astute sense of the political context of the armaments business.

In the age of big business, imperialist rivalries and naval arms races that followed, Elswick entered its period of national and international greatness. Between 1900 and 1914, the company was buffeted on the one hand by external political and competitive pressures and on the other by the internal difficulties of passing on the leadership of the firm to a new generation of managers. These managerial issues and competitive challenges have been seen as evidence of entrepreneurial decline. Nevertheless, all these problems were resolved, and in 1914 Elswick remained an industrial empire of considerable importance to the ability of Britain to achieve its foreign policy and defence objectives. Elswick played a significant role in re-establishing the supremacy of the Royal Navy against the navies of France and Russia after 1889, and in retaining that dominance of sea power against Germany after 1908.

Theories, models, approaches

The story outlined above will be told in detail with either direct or implicit allusions to several relevant theories and models which can unveil broader meanings and patterns to the narrative. However, the size and scope of Armstrong's activities make any single theory or model inadequate to capture the whole of the story. Historians of technology, politics, business, entrepreneurship and economic growth, and military or naval historians, will all find some episodes relevant to their particular theoretical interests. For example, historians of technology will seek – and find – evidence about the influence of non-technological factors in the design of Armstrong's gun, and the political process, as distinct from some purely objective technological measure, by which Armstrong's new gun design was proclaimed the best among all competitors. The battle to define technological superiority was essentially political, a question of debate and argument using skills which, on the surface at least, had nothing to do with engineering itself. The great Armstrong–Whitworth competition over what constitutes technological superiority is discussed at length in this study.

Business historians will no doubt attempt to measure the entrepreneurial performance of Armstrong and the other senior managers of the company. They will want to see Armstrong continuing technological innovation, modernising his company into a public corporation, seizing all opportunities to expand through mergers and acquisitions, integrating gun production forward into ship building and backward into steel refining and even coal mining, selling his wares through an efficient and aggressive marketing plan; and they will want to see the company recruiting the top professional managers to carry the company into the future. Finally, business historians will want to see the company's balance sheets and measure how efficiently its executives allocated resources according to changing market conditions. All these excellent questions are examined, but, as one might expect, there are many contradictions and complexities involved in answering questions about technological performance, and conclusions involve subjective assumptions and judgments which are open to question and debate.

Social and cultural historians will want to know if Armstrong carefully watched over his business daily, or if, once his personal fortune was made, he fled the noise and grease of the factory for the comforts and status of the landed aristocracy, with a great estate and a seat in the House of Lords. He did buy a great estate and he did sit in the House of Lords, but without cost to the business. Indeed, estate and ennoblement raised Armstrong's public profile to the great benefit of the company. Other historians will focus on how well Elswick competed in global markets. Did he know his customers well, maintain close contact with them and sell them the kinds of armaments they wanted, and provide financing when needed, or did he offer them only what he happened to manufacture, with a take-it-or-leave-it attitude? Actually, the greatest marketing problem was customers who did not know what kind of armaments they should acquire. Some business historians may be more interested in the relationship between the company's organisational structure and its business strategy; others in its ability to innovate or respond to new market conditions. Economic historians will be interested in the place of Armstrong and the armaments industry as a whole and its impact on the performance of the British economy as a whole. Was it a leading sector with positive effects on industrial growth in general, or was it an unproductive drain on scarce resources? Political and naval historians will no doubt focus on the nature of the relationship between Armstrong's private arms company and the British state, and in particular on the old 'merchants of death' question.

Social historians might insist that Armstrong's individual role was of secondary importance when set against the great social and economic forces of history which ultimately determined the fate of his business. That is so at the most general level, but the closer one looks the less it seems that Armstrong's innovations were simply and inevitably a result of the growing global demand for weapons. The contribution of Armstrong in bringing about that global demand for new guns and warships is another part of that historical process that needs emphasising. He provided states with armaments of unprecedented power and destructive capacity for their military and naval forces. His technological innovations and industrial enterprise changed national politics and altered the global balance of power. Men like Armstrong and Krupp did not overthrow the landed aristocracy and acquire political power directly, but their engineering enterprises clearly represent a new historical era to which all social classes and states had to adapt.

Finally, economic, international and cultural historians may read this work for the light it throws on the militarisation of British industry and culture which began around 1840, accelerated in the 1850s, declined in the 1860s and 1870s, but returned with a vengeance in the 1880s. Victorian militarisation was a complex historical process embedded in international rivalries and Victorian intellectual culture but, as one leading historian of this era rightly says, 'Without industrialization modern militarization is inconceivable.'[2]

Victorian Britain may have been the nation most satisfied with its international position and hence the least militaristic, but it was a leading force in applying industrial knowledge to military and naval power. Why this was so is an intriguing historical question which takes one beyond the scope of this work, but certainly Sir William Armstrong was Britain's pre-eminent military engineer and entrepreneur, and it was he, more than any other single individual, who militarised the Victorian industrial economy and provided the material basis for the militarisation of Victorian culture.

All these general themes and theories are in the Armstrong story, and to privilege one as the unifying theme would be to distort or deny the reality of the others. No one theory can support the narrative history of this extraordinary company, and the narrative history of this company cannot do anything to resolve the theoretical questions listed above. As one economic historian has concluded, 'most theories are at best partial'.[3] That is, no single theory can account for all the various developments and decisions in such a large entity as Elswick over a 50-year period. The full story can only be captured by several theories and models, and even then there remains a gap between theory and reality. Perhaps the concept of entrepreneurship comes closest to a grand theory, and much of the book is about entrepreneurship in one form or other, but entrepreneurship itself is an illusive idea involving all sorts of assumptions and values that cannot be awarded universal application.

This book attempts to confront several historical categories of analysis: historians of technology may see the sections on foreign sales as irrelevant; economic historians may consider the sections on inventing the Armstrong gun as a distraction; business historians may look upon the political and strategic contexts as secondary; finally, military and naval historians will no doubt find the business elements of less interest. But the overriding point is that all of these lines of analysis were important in the growth and development of the company, and to privilege one of them would be to leave only a partial account and a much distorted historical account.

For example, technological innovation is both distinct from and associated with entrepreneurship. Thus the book begins with technology and moves towards entrepreneurship, but it also shows that Armstrong's strong sense of applying technology to solve practical problems directed the kinds of technological innovations with which he concerned himself. Furthermore, while technological history has its own autonomy, that independence is relative, and affected by the non-technological context. Thus the development of Armstrong's artillery and the establishment of his armaments company are inseparable from the external world of war, British foreign policy, defence strategy and parliamentary politics. Those non-technological factors had much to do with the technological revolution brought about by Armstrong's new gun, but so too did Armstrong's new gun shape those external processes.

Part I examines Armstrong's mediating role in the interaction of various political, bureaucratic and social forces – members of the British Cabinet and Parliament, War Office officials, the community of engineers, newspapers – with technological possibilities at the time to transform an idea for solving a specific engineering problem into a breakthrough technology. Determining how exactly these different forces came together and shaped and reshaped each other is the aim of this contextual history.

Models of technological innovation

Joseph Schumpeter developed the first analytical model of technological innovation in the 1930s.[4] His model of technological change remains influential, as does his theory that, in a capitalist society, economic growth is driven by innovations in the factors of production. Schumpeter divided technological change into three relatively distinct activities: invention, innovation and diffusion, where invention is the development into practical reality of an idea for a new product, service, market or source of raw materials, or to improve significantly an existing one; innovation the commercialisation of that invention; and diffusion its successful deployment into the marketplace and its adoption or adaptation by commercial competitors. Schumpeter's model is sufficiently wide to include the practical realisation of existing knowledge by the creative application of known techniques and skills, to qualify as a true innovation. The inventor is one who combines input factors in new ways, or solves a pivotal technological problem. If he makes that invention a commercial success, he is a successful entrepreneur. Technological innovation, entrepreneurship and the broad social context are part of the same capitalistic process, but they also have their own autonomy. Technological problems are often those which are so defined by external, non-technological factors. So too the process by which one inventor's solution is deemed the best can be decided by an inventor's ability to mount an effective campaign in support of his particular innovation which convinces buyers that the particular characteristics of his invention are what they need. This is the social constructionist model of invention in which the precise design of a technology is not self-evidently superior, but gains that status by winning an argument.

A recent historian of technology has modified Schumpeter's model. Thomas Hughes separates invention from development and merges diffusion with innovation, leaving a three-stage model of invention, development and innovation. Here an invention is not an actual solution, but an idea which may solve a pressing technological problem on which many are working, and which, if successful, clears the way for a general advance along a broad technological front. Hughes uses a military analogy, referring to the blockage as a 'reverse salient along an advancing technological front'.[5] This idea too is parallel to Schumpeter's model, in which clusters of related technologies advance by sudden leaps. Stage two is development in which the proposed solution is tested, redesigned and tested again. This is the expensive stage, where great ideas fail because they do not solve the problem, or lack financial support. Innovation, the third stage in the Hughes model, is the process of moving the new technology from the laboratory and testing range to the commercial marketplace, and further into the general social and cultural context of the time. In this stage, business, political, legal, financial and cultural factors become more important, and those entrepreneurial skills take over from the purely technological. These entrepreneurial skills are needed to develop even a truly 'breakthrough' invention into an entire system upon which a society comes to depend in one way or another: a railway system, a military–industrial armaments complex or an electrical power grid. In this momentum stage a technology becomes a new factor in everyday political and cultural life, and the early phase, in which the external context exerts much influence over the design of the technology, is reversed and the technological system of the later stage exerts greater and greater influence over the political and economic contexts. New technological systems upset old systems, affect economic development and national security, and thus enter the political arena as a new weapon in the struggle for power. In the case of Armstrong, the early design stage, in which the context determined the design of his gun, shifted into the momentum stage in 1859 when its design brought about a reversal in British defence policy, from one that depended on the navy to one that depended on land fortifications.

The big question for historians of technology, then, is how non-technological, contextual factors – from the size of demand, available finances for research, wage rates and strength of competition to the political and cultural milieu – shape technological development, not merely by promoting one kind of technology over another (armaments over medical technology), but by defining what is the most efficient design. For example, Armstrong's attention was drawn to armaments technology by newspaper reports on the difficulties encountered in moving the very heavy cast-iron artillery pieces about the battlefield at Inkerman in the Crimean War. The technological problem was real enough: mid-nineteenth-century artillery materials (bronze or iron) and design (smooth bore, muzzle loaded) were virtually the same as those of the seventeenth century. Armstrong was only one of many engineers who sought to break this 'reverse salient along an advancing technological front' by inventing a rifled gun. But the problem was only made known to Armstrong by four external events: the Crimean War, the weather, the geography and the innovative decision of *The Times* to print detailed reports directly from the battlefield. As we shall see, these external, non-technological factors determined the design that Armstrong chose, and that design determined the materials from which he built his new gun in 1854. Similarly, non-technical factors played an important role in proclaiming Armstrong's design more efficient (at sinking ironclad warships) than the one made by Armstrong's greatest competitor, Sir Joseph Whitworth.

Technology and its external context are a major theme of the Armstrong story, and the discussion above reflects what historians of technology call a social interpretation to technology, a model of technological change in which the non-technological context plays a leading role, not only in directing the attention of engineers to a particular technological problem, but in determining to a great extent the specific design of the innovation. Strong social constructionists discard traditional 'objective' concepts such as the best or most efficient technology, and emphasise the 'subjective' social or political factors which define what the best technology should do exactly. From the social constructionist perspective, 'relevant social groups' or 'design constituencies' involved in the innovation process measure the efficiency of a particular design with a 'flexible interpretation' which allows for more than one design to be deemed superior. That is, a particular technological design not only is determined by the technological context, a step in a logical predetermined evolution of technology that is beyond human control, but also expresses the creative predispositions of its inventor and the broader social and political context. During the design stage, the technological design fluctuates and is open to contextual influences and change, a theoretical proposition investigated in the early chapters of this book.

Once a particular innovation has been proclaimed technologically more efficient than its competitors, it acquires a reputation and a momentum which speeds its integration into the broader social, political and economic context, and begins to exert an influence upon that everyday world. The balance of power between design and context, which had favoured context in the design stage of innovation, shifts in the momentum stage to favour the design in the momentum stage. While new technological designs are shaped by the external non-technological context, and as resistance by those who oppose the technology is overcome and the innovation accepted by the outside world, that balance of power is altered in favour of the technology. As more and more people come to depend upon it in one way or another, it becomes economically and politically difficult to change radically or to dismantle. In this momentum stage the impact of technology on the context outweighs the influence of the context on technological design.

Models of technological innovation are essential tools to guide research. However, as research historians are aware, records are not kept for the convenience of historians, and gaps and contradictions always remain between the model or theory and the evidence. Indeed, models and theories are out of fashion at the moment, partly because they have been overly employed and exaggerated, often in the heat of debates about the contemporary world, but also to mask the lack of evidence or a propensity to seek unity in the past. Nevertheless, models and theories of technological change are indispensable to understanding the emergence of modern armaments in the nineteenth century. Technology has its own processes, but those processes are shaped and directed by political decisions about which technologies and strategies will become policy. In a crisis the cause-and-effect relationship becomes fuzzy, relative autonomies become difficult to see clearly.

Finding the balance between these two statements is a central problem which this work seeks to find, or at least it tries to indicate that some balance of power between the two did exist. For example, there is no doubt that, between 1820 and 1860, Anglo-French diplomatic relations and mutual ambitions and fears played a role in the early use of steam engines and iron hulls in warships. Yet these non-technological contextual forces were not sufficient causes in themselves to bring about the technological revolution in military technology. Two other factors played a significant role: the internal history of technology on the one hand, and the commercialisation (and global proliferation) of armaments on the other. This convergence of technology and commerce, which was the core of the military revolution of the nineteenth century, was a convergence in which Sir William Armstrong played a leading role. This phase of the company history began in 1863 and, by 1890, the industrial empire which he built up had become an essential part of the British state and society. Between 1890 and 1914, naval armaments and the naval armaments business became more resistant to external factors. The balance of power between the armaments industry on the one hand, and politics, culture and diplomacy on the other, shifted to favour armaments.

The production of naval armaments acquired a strategic, political, economic and cultural momentum on a global scale by the turn of the century. Management of the huge industrial armaments companies occupied an important place in the vast military–industrial complexes that arose as the momentum of naval armaments industry replaced technological innovation itself as the driving force of the business.

Here too one cannot separate completely the internal managerial or entrepreneurial activities inside the company from the external context. As with the technological innovations of 1854–80, the managerial structures and styles of 1890–1914 that succeeded were those which best moved in accordance with the great contextual, political and international forces outside the boardroom.

Thus Armstrong's story is entwined with other relatively autonomous stories of technological innovation, business growth, parliamentary politics and international relations. Each of these forces moved by its own historical propellants, but each also was determined by the general context. In such a complex network, with ambiguous and overlapping borders, motives and actions, changes of course must be placed within a concrete context and allowances made for contingencies. Big theories about industrial capitalism and entrepreneurship are prone to empirical uncertainties and the historian must be content to strike some kind of balance between the evidence available and the scope of the interpretive argument.

A contextual approach to technological history attempts to find the balance between two extremes: the traditional view which sees technological change taking place without influence from outside social or political forces, and the opposite, post-modern claim that external non-technological forces determine technological history.[6] Certainly, for Armstrong, the external world could open or close opportunities. Disaster could follow unexpected success. For example, in 1859, an invasion panic in England and an Anglo-French naval arms race which propelled Armstrong into national prominence and launched his armaments business was followed in 1863 by financial collapse when that panic and arms race ended and the British government ended its contract with the new firm. Context: the market, international relations and parliamentary politics brought the company down in 1863. It would be far too unfair to argue that Armstrong's first attempt at the armaments business (as distinct from armaments technology) was evidence of his entrepreneurial failings. Indeed, the fact that he recovered and redirected the company into the world armaments market suggests a bold (that is, risky) entrepreneurial foray into the unknown. The incident shows that the political context is crucial to the Armstrong story and not merely background. The 1863 disaster contained two lessons for the company which it did not always remember: depending entirely on the British Admiralty could bring unprecedented profits, but those profits could also end suddenly.

Secondly, the company had to keep a close eye on the ebb and flow of the many contextual forces which determined its fate. British parliamentary politics, relations with war offices and admiralties, debates about the strategic implications of particular warship designs, arms races and the general international balance of power all interacted with the actions taken by the firm as it pursued its technological and business goals. These links between the internal history of the company and the external world shape the analytical aspects of the book and structure the basic story on what we may call a contextual narrative. This is the scholarly element of the work, the product of archival research which seeks to answer as objectively as possible basic questions of what exactly happened, in what precise sequence, who were important participants and what exactly they contributed to the outcome; and, finally, how these events affected both the company and its external context. This interaction between the internal managerial history of the company and the evolving external context is a major theme of the work which sets it apart from most business histories, including those on Armstrong, which tend to be overly internal. Indeed, one of the assumptions here is that business history must expand its horizon and recognise that the borders traditionally drawn between the internal developments of a business and the external world are artificial and confine business history. This book attempts to write the history not only from the inside of the company, but also from outside perspectives. A contextual and multi-perspective approach opens business history to much richer lines of investigation and provides a fuller understanding of business as part of a broad historical process.[7]

In Part II the focus shifts from technological innovation and the single market of the British navy to building an armaments business for global markets. Here the external political and competitive forces were challenges that the company learned to confront as they appeared. But the world was becoming more belligerent just as Armstrong began to supply it with his revolutionary new guns and warships. Part II thus examines the internal strategies and the external contexts which together transformed the company into a vast global enterprise.

Part III returns to a major theme of Part I: the relationship between the armaments industry, and Elswick in particular, and the British state: the Cabinet, the War Office, the Admiralty and Parliament. The earlier relationship began in 1854 but was broken off in 1863. The second attempt at creating a 'military–industrial complex' after 1884 was far more successful. Part III brings to a climax the overall argument that technological innovation and access to political and bureaucratic decision makers was the foundation upon which the company grew into one of the world's leading industrial enterprises.

Until his death on 27 December 1900, Armstrong was the key figure in the firm's success, but others were also important. Armstrong gradually withdrew from the daily management of the firm, not because he sought escape, but simply to retire. Andrew Noble took over the active operations. Armstrong had dominated the company, but he was not a domineering manager or a superman marching through life solving great technological problems alone, managing men and materials, succeeding while all around failed.[8] His success was due in part to many friends and associates. One whose contribution was second only to Armstrong's was Stuart Rendel. These two men established and maintained close links with the British state and brought the company through a few difficult early years. In 1913, Rendel re-established greater access to the British state which had been lost following Armstrong's death in 1900, and carried out needed managerial reforms within the company. These two men constituted one of the great partnerships in business history. Like Boulton and Watt, or their contemporaries and late rivals, Albert and Tom Vickers, their different talents, personalities and backgrounds complemented one another. Armstrong guided the technological and business direction of the company while Rendel cultivated the firm's reputation and developed a worldwide network of men who provided access to governments, the firm's customers. At the same time, each was knowledgable about the other's speciality. When Armstrong and Rendel coordinated their activities on a specific objective they became a powerful force. Armstrong was the great engineer and businessman, Rendel the salesman and negotiator, the man who had the right friends and contacts to help get the company off to a strong start. Armstrong's natural desire to avoid conflict or controversy was balanced by the outgoing, combative Rendel. Armstrong lived in Newcastle, where the factory was located, and Rendel lived in London, where governments of the world sent their representatives.

In the final analysis there is no doubt that Sir William Armstrong was a very successful Victorian entrepreneur. Between 1862 and 1914, his armaments company was among the largest, most innovative, progressive and profitable companies in the world. The latest production techniques were employed, the importance of scientific research and development was preached and practised and capital investment remained high. The company did not hesitate to restructure itself into a public company in order to expand and it was a leader in the amalgamation movement which erupted towards the end of the century.

It has been argued that, because armaments are sold only to governments, armament manufacturers are a special group of industrialists, that the armaments business is by nature more concerned to develop new technologies based on science and research and that the continuous belligerence among nations in the late nineteenth century did not allow the armourer kings the periods of relaxation which their contemporary manufacturers of civilian goods enjoyed. National rivalries fuelled weapons and research and development with greater intensity than mere competition amongst capitalists. With its monopsonist position, says Clive Trebilcock, who has written extensively on this topic, 'the state procurement agency wielded enormous market power over its suppliers' and insisted that private manufacturers construct to the highest technical specifications. Any contractor who failed to satisfy these exacting demands could lose its only customer. Thus a routine of technological innovation was forced upon armaments producers by the state.[9] This model does not fit the facts of Armstrong's relationship with the British state. His greatest technological innovations were not simply a consequence of demand; indeed, he struggled constantly to prompt the War Office to accept new technologies.

Armstrong's relationship with governments was complex: his prime goal was to gain access to the political decision makers, but the bureaucracies had great power to sway or shape policy and were competitors if their interests were at stake. The Royal Gun Factory at Woolwich did not welcome Armstrong's intrusion into their domain. The Admiralty was sometimes an ally, sometimes not. Circumstances were always important to the state of relations at any given time. Armstrong did not have 'a' relationship with 'the' British government but several kinds of relationships with different arms of the British state, all of which fluctuated as circumstances changed. A definition of the British state which reflects this approach is that offered by Ralph Miliband: the state is not a thing as such, he says, but it stands for 'a number of particular institutions which, together, constitute its reality, and which interact as parts of what may be called the state system'.[10] How Armstrong negotiated with state systems is a unifying theme to this history.

Notes

1. On the facts, theories and models in history, see G.R. Elton, *The Practice of History* (London: Fontana, 1969) and Richard J. Evans, *In Defence of History* (London: Granta, 1997).
2. Geoffrey Best, 'The Militarization of European Society, 1870–1914', in John R. Gillis (ed.), *The Militarization of the Western World* (New Brunswick and London: Rutgers University Press, 1989), p. 14.
3. Charles Kindleberger, *Economic Growth in France and Britain, 1851–1950* (Cambridge, Mass.: Harvard University Press, 1964), p. 324.
4. Schumpeter explains his theory of innovations in several of his works: *The Theory of Economic Development* (Cambridge, Mass.: Harvard University Press, 1934), *Business Cycles* (New York: McGraw-Hill, 1939), vol. 1, and 'The Instability of Capitalism', *The Economic Journal* 38 (1928), pp. 527–42.
5. Hughes explains and develops his model of technological change in several works: *Networks of Power: Electrification in Western Society, 1880–1930*

(Baltimore: Johns Hopkins University Press, 1983), a contextual history of electrical power generation and distribution systems in Germany, Britain and the United States; *American Genesis: A Century of Technological Invention and Technological Enthusiasm* (Harmondsworth: Penguin, 1989); 'Technological Momentum', in Merritt Roe Smith and Leo Marx (eds), *Does Technology Drive History? The Dilemma of Technological Determinism* (Cambridge, Mass.: MIT Press, 1995), pp. 101–13; 'Technological Momentum: Hydrogenation in Germany, 1900–1933', *Past and Present* (August 1969), pp. 106–32.

6. Among the many studies linking technological history to social, economic and political history which are relevant to this work on Armstrong are two books by David Noble: *America By Design* (New York: Knopf, 1977) and *Forces of Production* (New York: Knopf, 1984). For a bibliographic overview, see John Staudenmaier, 'Recent Trends in the History of Technology', *American Historical Review* 95 (1990), pp. 115–25, and *Technology's Storytellers: Reweaving the Human Fabric* (Cambridge, Mass.: MIT Press, 1985); Smith and Marx, *Does Technology Drive History?* For a post-modern approach to technology, see W.E. Bijker et al. (eds), *The Social Construction of Technological Systems: New Directions in the Sociology and History of Technology* (London: MIT Press, 1987). Traditional histories of technology include the eight volumes edited by Charles Singer et al., *A History of Technology* (Oxford: Clarendon Press, 1954–84), and Donald Cardwell, *The Fontana History of Technology* (London: Fontana Press, 1994).
7. Two examples of this approach are Peter Payne, *Colvilles and the Scottish Steel Industry* (Oxford: Clarendon Press, 1979) and Paul Kennedy, *The Realities Behind Diplomacy: Background Influences on British External Policy, 1865–1980* (London: Fontana, 1981). Charles Kindleberger argues that theories are necessary and useful but are by nature guides which necessarily exclude much about the complex realities of life. I agree, but would emphasise that theories exist even in the minds of historians who deny they exist. Theory is embedded in any thought or statement.
8. For an example of this kind of history, see Clive Trebilcock, *The Vickers Brothers: Armaments and Enterprise 1854–1914* (London: Europa, 1977). Other works on the company are David Dougan, *The Great Gunmaker: The Story of Lord Armstrong* (Newcastle: Graham, 1971) and Peter McKenzie, *W.G. Armstrong: The Life and Times of Sir William George Armstrong* (Morpeth: Longhirst, 1983) which are adequate introductions and surveys but of light scholarly weight. The few chapters on Armstrong in J.D. Scott, *Vickers: A History* (Weidenfeld & Nicolson, 1962) have remained the source upon which historians of British armaments industry have relied for information about the company. All these books are 'official' accounts. An excellent book is Kenneth Warren, *Armstrongs of Elswick: Growth in Engineering and Armaments to the Merger with Vickers* (Basingstoke: Macmillan, 1989), a totally internalist examination of Elswick's entrepreneurship.
9. Trebilcock, *The Vickers Brothers*, pp. 2–3.
10. Ralph Miliband, *The State in Capitalist Society: The Analysis of the Western System of Power* (London: Quartet, 1973), p. 46.

PART I

INVENTING THE ARMSTRONG GUN, 1854–65

CHAPTER ONE

War, friends and competitors

I

On a December day in 1854, William Armstrong, a successful businessman and engineer from Newcastle, was ushered into the office of the Duke of Newcastle, the British Secretary of State for War. Armstrong carried drawings of a new cannon which he believed constituted a significant advance over all existing artillery. The War Office, the department of the army responsible for supplying both the British army and the navy with their guns, had not responded to the many civilian engineers who had submitted designs for new weapons, and Armstrong thus sought the assurance from Newcastle that, once he had developed his idea into an actual gun, the War Office would give it a fair trial. Access to the duke had been cleared for Armstrong by his good friend James Rendel,[1] the navy's chief civil engineer. Rendel had asked the First Lord of the Admiralty, Sir James Graham, to arrange the meeting between Armstrong and the duke.[2] Thus helpful friends in high places gave Armstrong an enormous social advantage over other inventors clamouring for the government to give attention to their ideas.

Newcastle listened to Armstrong eagerly. The Crimean War had barely begun but was already turning into a military and hence political disaster for the government.[3] The most obvious problem was the poor supply lines to the troops, administrative confusion and incompetent military leadership, but there were three related technical problems: the range of British smooth-bored guns was shorter than the rifled small arms of the Russians, they had not the power to break the walls of the huge fortress at Sebastopol and, thirdly, they were of such bulk and weight that it was extremely difficult to manoeuvre bronze and iron smooth-bored cannon quickly around the muddy hills of the Crimean battlefields.[4] The problem was revealed during the Battle of Inkerman on 5 November 1854. The Russians attacked and Lord Raglan, the Commander-in-Chief of Britain's expeditionary force, ordered two 18-pounders brought up to defend their position. No draught horses were available, however, and it required 150 soldiers and eight supervising officers three hours to haul the two-ton guns into position.[5] The Russian attack at Inkerman was finally beaten back, but only upon the arrival of allied infantry armed with French Minié rifles and fierce hand-to-hand fighting which cost 2500 British casualties.

The invention by the London *Times* of publishing reports sent from the battlefield had a tremendous impact in England. It captured the interest and imaginations of its readers, and made the war an immediate event in their lives, which provoked reactions in many quarters. It drew Armstrong's attention to the state of armaments engineering as compared to civilian accomplishments, and presented a new engineering opportunity for Armstrong to pursue. The political scandal of the war centred on the administrative chaos, the breakdown of the supply lines and the costly tactical mistakes made by the government at home and by generals on the battlefields. Armstrong, however, concentrated on the outdated British artillery. He read the newspaper report on Inkerman at breakfast in the London home of James Rendel, with whom he stayed whenever business brought him down from Newcastle. Rendel raged against British military engineers who had not improved artillery since the seventeenth century. Compared to civil engineers like himself and mechanical engineers like his friend Armstrong, he said, military engineers were an apathetic and backward lot who remained wedded to heavy cast-iron material which was too brittle to be rifled. Wrought iron was lighter and softer and a gun forged of it could be rifled as small-arm gun makers had proved. Rendel proclaimed Armstrong the man to bring British artillery into the nineteenth century and pressed his friend to turn his skills to artillery.

Armstrong accepted the challenge as an interesting engineering problem, not as a new business adventure. This distinction is important in that it reminds us that technological innovations are not always motivated by profits. Armstrong's hydraulic crane and mechanical engineering business was prospering, and Armstrong was open to new areas of engineering. He did not envision moving into the armaments business in 1854, and indeed entered the business reluctantly in 1863. He entered the armaments business from a different context and for very different reasons than did the Vickers brothers in 1889. They switched because the demand for rails had dried up, and the demand for naval armour was on the rise. In 1854, Armstrong looked at armaments as an engineering challenge. Business considerations came later. Which of these motives is the more entrepreneurial? From which come truly innovative changes in technology? Armstrong's entrepreneurship straddled both technology and business, but his move into armaments began with an engineering problem which had haunted military engineers for centuries.

II

The history of gunpowder weapons in Europe begins in the thirteenth and fourteenth centuries. Medieval gun makers had tried to build guns of wrought iron but found them weaker than bronze or, to a lesser degree, cast iron, and they returned to casting their heavy guns.[6] These heavy and crude mechanisms were dragged to the battle site by oxen and worked entirely by hand. Their range and accuracy were matters of luck as much as the skill of the gunner. These great iron monsters gave rise to the era of early modern siege warfare during which gun makers and military architects engaged in a see-saw battle of offence against defence.[7] This 'military revolution' of early modern Europe was not fully embraced in Britain. It remained 'a zone where the transformation in fortification and siege craft was incomplete, gradual and relatively tardy'.[8] English military architecture came to a virtual halt after the War of the Roses when, in the era of Tudor peace, the great landowners built comfortable manor houses to replace their fortified castles. When Henry VII built a string of forts to meet a French invasion, he refused to hire a European professional military architect and was only brought round by the reality of such an attempt in 1545.[9]

Artillery emerged as an effective weapon during the fifteenth and sixteenth centuries, when its size was reduced and mobility increased. Charles VIII's invasion of Italy in 1494–5 was successful in large part because his artillery, drawn in wheeled vehicles by horses instead of oxen, was never far behind his infantry. Mobile artillery allowed him to accomplish in a few hours what had once taken days.[10] In the seventeenth century, Gustavus Adolphus lightened his artillery with differing bronze alloys, sought to standardise it to three calibres and developed a rote-learned system of speedy loading which allowed his gunners to fire 20 rounds an hour, a rate which approached that of musketeers.[11] The French made further technological improvements during the eighteenth century. Their more precise boring techniques reduced the windage (the space between the shot and the inner wall of the barrel) and directed more of the explosive force against the shot. Less powder – and thus less explosive force – imparted the same muzzle velocity and hence artillery could be made with thinner walls. The weight of a French 4-pounder fell from 1300 pounds to 600 and the French 12-pounder fell from 2400 pounds to 1200.[12] Artillery had become an effective mobile battlefield weapon by the time of the French Revolution and played a major role in Napoleon's sweep across Europe.[13]

Throughout this period major developments in artillery were products of European inventors and military leaders. The English contributed little to either the technology or the tactics of artillery warfare. Gustavus Adolphus directed 90 pieces of artillery at an action in 1632, but in some battles of the English Civil War field guns were entirely absent. Many attempts to imitate European light and mobile 3-pounders ended in failure.[14]

Small-arm gun makers reclaimed some advantage for Britain after 1815 with the development of conical bullets, percussion caps and rifling. The range and accuracy of 'rifles' exceeded those of artillery and hence infantry could drive back the artillery. Infantry's advantage was enlarged further when Johann Nicholas von Dreyse invented the needle-gun in 1836 and opened the way to breech-loaded rifles with which soldiers could fire and load from crouched or prone positions. The French Minié muzzle-loaded bullet gave rifles a range of 1000 yards in the 1850s and it was followed by breech-loaders with ranges up to 1600 yards.[15] Artillery inventors attempted to adopt these new technologies in the 1830s and 1840s but failed to produce a practical rifled artillery piece with which to counter the small-arm rifle.[16] Inventors in America and France had made progress by the 1850s, but an artillery piece that combined the latest developments in small arms eluded them.

England seemed the least likely place for a breakthrough to occur. Britain was the first industrial nation, but, as Rendel remarked, her military engineers remained content with their guns and her civilian engineers were busy making steam engines, roads, railways and ships. Even Napoleon's triumphs had failed to generate new thinking about British field guns. The complacency among military authorities, bred by victory in 1815, was underscored by British middle-class faith that material progress and international trade were making war obsolete. The gentry who supplied the officers and the rural labourers who supplied the soldiers for the British army were the declining classes of the nineteenth century while the rising middle-class looked to the navy for whatever defensive needs Britain required.[17] Nevertheless, advances in metallurgy (the refining and fabricating of iron ore) and chemistry, the invention of the steam hammer and the development of high-temperature furnaces provided the technological knowledge and skills with which wrought iron could challenge the traditional gun-making metals. The technology was present; it needed an imaginative engineer to organise those resources in new ways.

III

Until the Crimean War, Armstrong kept busy building up his hydraulic engineering company and took no interest in military technology. Indeed, there is little in his family background to indicate that he would become an engineer of any kind. His father, the son of a Cumbria yeoman, moved in the late eighteenth century to Newcastle, where he worked his way up from a clerkship to a partnership in a corn-trading business. He was an active municipal reform politician in the 1830s and was elected mayor in 1850.[18] Born in 1810, young William grew up with all the middle-class advantages of the day: a successful father, a highly cultured mother, private schools, grammar school. His father wanted him to be a lawyer and young William was articled in a prominent Newcastle law firm, studied at Lincoln's Inn, returned to Newcastle in 1833, married, and began a promising career as a partner in Messrs Donkin, Stable & Armstrong, where he oversaw the legal interests of many important families and businesses. He was intelligent and hard-working and was without doubt on the way to a high place in the world of law and commerce, and perhaps politics.[19]

However, his early interest and natural aptitude for science and technology eventually led him to abandon law for engineering. At the age of six, he had constructed mechanical devices of such sophistication to impress guests, one of whom was Robert Stephenson, the famous engineer and Conservative Member of Parliament for Whitby. While he was practising law in the 1830s, Armstrong's attention was drawn to the study of electricity by the experience of a coal miner who received an electrical jolt when, with one hand on a boiler, he shoved the other into the jet of steam emitted by that boiler. Armstrong's inquiring mind pursued this phenomenon with experiments and the eventual invention of an electrical generator. Between 1841 and 1843, he published several papers on production of electricity by jet streams of water. As his boyhood mechanical devices had impressed Robert Stevenson, Armstrong's 'hydro-electric' machine was of much interest to Michael Faraday. In 1846, Armstrong produced a hydraulic crane, the basic mechanism which led to his immediate success as a crane manufacturer. His first hydraulic crane incorporated a mechanism which multiplied the pressure exerted on a piston inside a cylinder by the water pressure in local town water mains or special reservoirs. Armstrong built his first hydraulic crane on the Newcastle quay in 1846. He first convinced a receptive Newcastle City Council to use his hydraulic engine to supply Newcastle with fresh water, and the Whittle Dene Water Company was formed, with Armstrong the company secretary and Donkin, Stable & Armstrong its solicitors. Armstrong left that law firm to devote himself to his new adventure in technology and business. A year later he again combined technology, business and government when he received permission from the Newcastle Council to use his hydraulic cranes to load and unload ships, a rapidly growing activity in Britain at the time.

The business grew rapidly, a process which naturally brought more technological challenges to Armstrong. Since there was no practical external source of water pressure at New Holland, on the Humber estuary, he incorporated a steam engine into his hydraulic system which provided the necessary water pressure on the piston. This modification freed Armstrong's hydraulic system from requiring an external source of water pressure. It was this autonomous hydraulic system that allowed Armstrong's heavy guns to be manipulated. This was especially important for the very large guns on warships. Armstrong's hydraulic crane was thus the basis for his success in the late 1840s as a crane manufacturer, and his later, greater fame as an armaments king. Indeed, in 1890, when any competent military engineer could make a good naval gun, Armstrong's Elswick factory remained the only source of the high-quality naval gun mountings, without which the huge naval guns of the day could not be manipulated.[20]

James Rendel played an even more direct role in launching Armstrong's crane business in 1846 than he would in helping Armstrong to become an armaments manufacturer in 1854. Rendel ordered Armstrong cranes for Admiralty docks and was instrumental in obtaining a contract for the Liverpool docks for the Elswick Engine Works. As with naval armaments later, Admiralty orders greatly increased Armstrong's engineering reputation and prompted more sales. I.K. Brunel, one of England's most famous engineers, ordered Armstrong's hydraulic locomotive turntables for the new Paddington station.[21] Elswick hydraulic machinery became a tremendous business success. By 1850, Armstrong was a rich man with a national reputation as an inventive engineer.

Armstrong's interest in science and technology remained active throughout his life.[22] Despite armaments becoming the overwhelmingly dominant division at Elswick, the Engine Works continued to manufacture hydraulic and electric crane hoists, machinery to operate swing bridges (including the one on Tower Bridge), dock gates, sluices, capstans and all kinds of hydro-electric products.[23] In 1873, Armstrong travelled to Egypt, one of the first armaments customers, where he examined the possibility of building bridges across the Nile and wrote a book about his visit.[24] For Armstrong, business was in the service of technological innovation, an important but secondary means by which he could develop science and technology into useful products. Unlike the Vickers brothers in 1889, Armstrong did not enter the armaments business in 1859, or more pointedly in 1863, because of falling profits in his civilian engineering business. He was a cautious businessman, a quality which is often confused with poor entrepreneurship, but his engineering imagination and technological daring were second to none during a period when technological innovation was the foundation of business success. In the industrial age an engineer can fail in business however innovative his technology; but even the most astute businessman will certainly fail in the marketplace without a practical technology.

Armstrong was thus a Victorian engineer with wide scientific and technological interests, a typical participant of the heroic age of industrialisation, when great engineers sought to expand their technological knowledge and apply their engineering skills everywhere, and let the money fall where it may. He invented cranes to unload ships and then guns to sink them. He had a passionate interest in pushing out the frontiers of Victorian technology. Had he remained a civilian engineer he no doubt would have continued to make further major contributions, but Victorian engineering was on the threshold of a new era and many technological challenges and business opportunities lay before Armstrong in 1854. His crane business was established and profitable and he was ready to seek new technological borders to cross. Thus, when James Rendel challenged him to take up the artillery question, Armstrong did not resist. After breakfast the two men huddled over the dining table where Armstrong sketched a design of a wrought-iron rifle which he thought could be enlarged into a field gun. After further work and consultation he completed his designs in December. These were the plans that Armstrong carried into the chambers of the Duke of Newcastle.[25] Within six months, Armstrong would construct an artillery piece that would embody a new business, a new industry and a new era.

IV

The Crimean War revealed in a dramatic way how far artillery engineering had fallen behind small arms in both material and design, and precipitated tremendous interest in weapons amongst British engineers. The engineer who could enlarge a small-arm rifle into a heavy artillery piece would solve an age-old pivotal engineering problem, open up the entire field of military and naval engineering and be acknowledged as a great engineer. The Crimean War stirred patriotism with technology to create the context which attracted the attention of newspapers, engineers and governments to military engineering. After all, it had been only a few years since the Great Exhibition had proclaimed that all engineering problems would be solved. This is not to deny the profit motive was at work, only to insist that in 1854 armaments offered engineers an opportunity to solve an interesting engineering problem brought to their attention by the war, not from some internal momentum or logic within military engineering itself.

In 1854, artillery and ordnance entered a design phase in which many engineers worked on the same problem and the various ideas for a new gun were the subject of sharp debate and competition, and open to political and social influences. After 1880, military and naval engineering acquired a momentum which reversed its relationship between the social and political contexts and also pulled military and naval engineering ahead of civilian, thus reversing that ancient relationship. In the second half of the nineteenth century, military engineers took the lead and set new standards, developing new techniques for civilian engineers to follow.[26] The breakthrough in artillery technology achieved by Armstrong and a few others, such as Krupp, represents the beginning of a 'military revolution' of the industrial age. Armstrong's 1854 innovations in rifling, breech loading, shells, fuses and the all-important hydraulic carriage for aiming and firing began an historical process of merging civilian and military engineering, and laid the political foundation for the sprawling military–industrial complexes of the late nineteenth century.

Even more importantly, Armstrong initiated an historical trajectory of the commoditisation of weapons, and the concomitant militarisation of British engineering. After 1854, weapons attracted greater numbers of bright and ambitious engineers, who found the most interesting and challenging of engineering problems to overcome. This necessitated making arms production a capitalist enterprise to generate profits for further research and development.

The third step in this historical process came in the 1880s and 1890s, when British foreign and imperial policies, along with those of other states, began to take into account, to an extent unknown in the past, the impressive and rapidly evolving technology of armaments, and the increasing demands of a market-driven armaments business. Whereas in the 1850s and 1860s the political context determined the design of armaments, after the Naval Defence Act of 1889 the design of armaments exerted greater and greater influence over domestic and international politics.

The Crimean War, then, attracted the attention of many European engineers to military technology, and in particular the problem of rifling.[27] British inventors sent their plans and ideas to the War Office and anticipated a call from a grateful government anxious to support the development and production of their ideas. However, the army's procurement system was not designed to handle a large number of applicants and it was overwhelmed by the deluge. Between June 1854 and the autumn of 1855, the War Office was swamped by nearly a thousand proposals to increase the firepower of Britain's armed forces.[28]

Dissatisfaction among the inventors grew as the government failed to respond to their offers of help. Bashley Britten of Mandslay, Sons and Field's waited nine months for the government to act upon what he considered his proven idea.[29] Another inventor protested at the government's 'meaningless lithographed replies' to his proposals.[30] Various tactics were tried to hurry the process. Alexander Parkes announced a public demonstration of his shell.[31] Theophilus Blakely searched Constantinople and Balaclava for Lord Lyons to persuade him to try his gun.[32] Nothing worked: the Board of Ordnance seemed impervious. Failing to gain public support through the press, some inventors warned they would take their ideas to the 'general scientific world' gathered in Paris for the 1855 Exhibition. A few threatened to go to 'other Powers'.[33] In the name of technological progress, they said, inventors needed to have confidence that their efforts were recognised.[34] As this confidence waned, they accused the Ordnance Committee of incompetence. Soon they detected a more sinister reason for the delays: the Committee of Artillery Officers, the sub-committee of the War Office's Board of Ordnance which examined proposals and recommended which ones ought to pursued, was stealing their plans for itself.[35]

Sir John Burgoyne later explained the War Department's predicament to I.K. Brunel. Civilian inventors and would-be inventors, he said,

> send in the most outrageous propositions which the projectors, however, cling to with pertinacity, and call us bigots, narrow-minded, and fools, because we will not adopt them. You can have no idea of the quantity of wild and undigested propositions that are made to us. I can assure you that the fall of Sebastopol was quite a relief to me, personally, by putting an end to the continuous stream of projects with which I was inundated for taking Sebastopol at once ... The fact is, a sudden thought strikes a gentleman, and he considers the matter settled without taking any trouble to study and analyse it, and even men who may be very able in their own line fail in that way.[36]

Inventors were determined however, and as the Crimean campaign continued they gained enough public support to force the government to respond. Newcastle, citing 'the present circumstances of the country', disbanded the Committee of Artillery Officers on 8 January 1855 and replaced it with the Ordnance Select Committee, which included civilians.[37]

The War Office resisted this civilian penetration of its sanctuary. Raglan protested at the loss of his job as Master General of Ordnance, but was told that the public agitation from 'high and low' made it politically necessary.[38] General Ross thought that Newcastle was unaware of the 'real construction' of the old Committee and Major General Eaton insisted that 'appeals are always met in a spirit of fairness and justice to the inventor'. They were unaware of any complaints. General Cator protested that many inventions were ingenious but impracticable; many were similar to others that had already failed, but the great mass of them were 'utterly worthless'. The enlarged committee was cumbersome and inefficient, he said, and he begged Newcastle to leave artillery questions in the hands of military experts. Newcastle reminded Cator that many of the most eminent English artillerists, such as Lancaster and Nasmyth, were civilians.[39]

The make-up of the Ordnance Select Committee was overwhelmingly military and the inclusion of a few civilians merely made formal the practice of the old Artillery Officers Committee of seeking outside consultation from time to time.[40] General Ross soon realized that the new committee had greater power and autonomy.[41] Inventors still had to send their plans to the Ordnance Committee.[42] Delays continued and more questions were asked in Parliament.[43]

The most contentious aspect of the new committee was its expanded authority to develop its own guns as well as to evaluate the designs of others.[44] The government's arsenal at Woolwich was now in direct competition with the trade and yet had access to their secrets.[45] In short, Newcastle's reforms exacerbated rather than allayed the fears and suspicions of civilian inventors. General Peel dropped civilians from the Ordnance Committee in 1859.[46] By that time, Armstrong had swept aside all competition and emerged as Britain's foremost gun maker.

V

Armstrong had the help of two of England's most eminent engineers, Isambard Kingdom Brunel and James Nasmyth. They had been working on artillery questions before Armstrong. Two months before Inkerman, Brunel had designed a large gun.[47] Two years before that Nasmyth, responding to the invasion panic that swept over England when the Second French Empire was proclaimed on 2 December 1852, published details of a giant shell to be fired from a 600-ton mortar. His shell, he said, would destroy, 'by one masterly blow, the largest ship of an invading enemy'.[48] The invasion scare of 1852–3 faded away, however. The Prime Minister, Lord Aberdeen, had no desire to encourage it and even the bellicose Palmerston wanted good relations with France. Sidney Herbert, the Secretary at War, thought a French invasion unlikely.[49] Thus Nasmyth's project did not receive government financial support.

At the outbreak of the Crimean War, the government quickly increased its production of small arms. A government cartridge factory was built and Joseph Whitworth, who knew nothing about ordnance but everything about precision manufacture, was given nearly £13000 to develop rifling machinery for small arms. Five others received lesser amounts for their projects. But after Inkerman the government could not ignore the artillery question. Men like Nasmyth, Brunel and Armstrong were needed and the government opened its Treasury to them. Over the next five years it granted nearly £60 000 to 23 civilian artillery inventors working on 25 projects. Rifling experiments were given strong support. Charles Lancaster received £10 000 to develop his oval-bored barrels and Whitworth another £4247 to rifle cannons hexagonally. The largest single grant, £11 807, went to Mallet to build his large mortars. Hale was given £7810 for rocket research (explosive shells or bombs fired with high trajectories from mortars).

Armstrong received £7219, by far the largest amount of those working on rifled breech-loaders.[50] Since Nasmyth was in effect Armstrong's partner, his £3000, awarded in early December 1854, 'to test the applicability of malleable iron to large ordnance', should be added to the Armstrong amount.[51] Brunel did not apply for a grant. He had become immersed in the construction of his new steam ship, the *Great Eastern*, and had little time to work on artillery. He therefore asked Armstrong to construct a gun on the design he had developed. As Armstrong explained a few years after Brunel's death:

> Soon after I began to direct my attention to artillery the late Mr. Brunel consulted me upon a plan of constructing cannon by enveloping an internal tube with wire. In theory the idea was perfectly sound, but there were various practical difficulties and objections to be removed. After fully considering these Mr. Brunel determined to have a gun made upon this plan at his own expense, and I undertook to carry out the experiment for him.[52]

The roots of what became known as the Armstrong gun lay in the coordinated effort of Armstrong, Brunel and Nasmyth.[53] Nasmyth worked on shell designs and Brunel and Armstrong on the gun. The collaboration of two of the greatest nineteenth-century engineers was of enormous benefit to Armstrong, but Brunel and Nasmyth are part of only the first stage of the invention of the Armstrong gun. Brunel and Nasmyth no doubt contributed to his success, but, in the end, Armstrong deserves the credit for the gun that carried his name. When Armstrong's gun gained general recognition in 1858, Brunel was killing himself on the *Great Eastern* (he died in 1859) and Nasmyth had moved on to other things. Armstrong's role of turning ideas into practical reality had been the key to his success in building his hydraulic cranes and he would repeat that success with artillery.[54]

The gun Armstrong developed had a number of remarkable features, but four stand out. The first was that it was a rifled artillery piece. This characteristic followed from the fundamental requirement that the gun, like the small-arm rifle, have long range and be accurate. Range was directly related to muzzle velocity, but also to the shape and spin of the shell. An elongated shell has greater range than a round shot, even if the latter is fired with a higher initial (or muzzle) velocity than the former, since its shape sharply reduces air resistance. Accuracy, the sureness of flight path, was increased by imparting a spin on the projectile during its passage through the grooved (rifled) barrel. The spin was also necessary to keep the elongated shell point-forward during flight and allow the percussion fuse in the nose to strike the target.

The second important feature of the Armstrong gun was that it was loaded at the breech, a practical consequence of rifling.[55] Medieval artillerists had made breech-loaders but they were forced to abandon them (as they had abandoned wrought iron) when they failed to construct a tight breech mechanism strong enough to withstand the explosion of the charge. Breech-loading too had to await the advances in metallurgy and precision mechanical engineering made during the Industrial Revolution. In Armstrong's gun the shell was loaded through a hollow breech-screw and into the chamber, followed by a cylindrical bag of gunpowder. A steel block, or vent-piece, was dropped into the chamber and pressed with the breech screw air-tight against the opening of the bore. The shells were coated with soft lead such that their diameter slightly exceeded the calibre of the gun and engaged the rifling immediately upon firing. The gun fired segment, common or incendiary shells. A time fuse could also be used to have the shell explode just before it struck, or just after it had penetrated the target.[56] The gun combined virtually all the properties that make up an effective artillery piece. Armstrong mounted his technological *tour de force* on a carriage which directed the recoil up an incline slide from which gravity returned it to firing position.

The Armstrong gun was a technological masterpiece and set new standards in gun making, but it was the third and fourth features of the Armstrong gun that made it revolutionary. It was Armstrong's method of constructing the gun out of wrought iron that cleared the way for the rapid evolution of heavy guns into chimneys of unprecedented destructive power in the second half of the nineteenth century. The characteristic which makes the Armstrong gun a turning point in the history of armaments was its 'built-up' or 'coil' method of construction, and it was this distinctive feature of the gun that was meant when it was referred to as the 'Armstrong gun' or a gun built on the 'Armstrong system'. This method was possible only with wrought iron, the fourth remarkable characteristic of the Armstrong gun. It was a simple procedure but, as Armstrong repeated many times, carrying it out successfully was a major achievement. Armstrong's built-up or coil gun was made in layers of wrought-iron tubes. Wrought-iron bars were made into a solid tube by coiling them around a mandrel and welding the edges with heat and pressure.[57] A second tube was made of smaller inner diameter than the first. This tube was then heated until its inner diameter had expanded enough to slip over the cold first tube. As the second tube cooled it shrank tightly onto the inside tube, thereby compressing it. Armstrong had carefully experimented and calculated this procedure so that the second, outside tube was prevented by the inside tube from shrinking back to its original cold diameter. Thus an even pressure was maintained throughout the walls of the gun which was directed in the opposite direction to the latitudinal forces exerted on the barrel when the gun was fired. Without this pressure the outer layers of the barrel would contribute less and less to the strength of the gun.[58] The procedure was repeated with a third tube and the gun was 'built up' in this manner. Thicker tubes were used near the breech, where the pressures from the detonation of the charge were greatest (hence the telescopic shape of the early Armstrong guns).[59]

Armstrong built a 3-pounder in this way over the winter of 1854–5 and submitted it to the War Office Ordnance Committee in July 1855. It was a small gun but its performance was extraordinary. From 2000 yards its 3-pound projectile penetrated over two feet of elm timber backed by a thin cast iron plate. Its accuracy, range and power encouraged the War Office to ask Armstrong to enlarge it into a 5-pounder. On 30 December 1856, Armstrong tried the larger gun successfully at Newcastle and in January 1857 it was purchased by the War Office. Only a year later, on 26 January 1858, an Armstrong 18-pounder was tested against a Service 32-pounder muzzle-loader. The technological leap represented by this gun became clear at this point. The Secretary of State for War, Lord Panmure, was 'clearly of opinion that this is a most valuable contribution to our Army'.[60] The 'built-up' rifled piece represented a quantum leap in the proficiency of artillery. Armstrong was given an order for two more 18-pounders, one 12-pounder and 400 projectiles. Shortly thereafter the new Derby government ordered 100 guns and guaranteed the £12 000 investment by Armstrong and his partners to build a gun factory beside his crane works at Elswick, on the outskirts of Newcastle. This was a significant order and there seemed little doubt that more would follow.

Armstrong, by applying a cluster of small-arm technologies, had developed an artillery piece which resolved most of the technological problems that had frustrated artillery makers since the fifteenth century. Behind him stood many men of mechanical genius upon whose work he built, not only gun makers but all those who had contributed to the tremendous advances made in the iron industry during the Industrial

Note: Note the telescopic 'built-up' method of construction.

Figure 1.1 Armstrong 6-pounder gun, Canadian War Museum, Ottawa

Revolution. How much did Armstrong owe to James Nasmyth for both his work on shells and his steam hammer, without which the coils could not have been fused into tubes? How much did he owe to Dr Barlow of the University of Cambridge, who explained to Armstrong the mathematics of the stresses inside the walls of a gun when it was fired? What was Armstrong's particular contribution to the gun that bore his name? These questions cannot be answered in any clear way. About the best we can say is that Armstrong was an engineering genius who was born on the right side of the Industrial Revolution. In the 1850s, however, the question of credit for an invention was a live one and fortunes might rest on its answer. Some gave Armstrong all the credit, others gave him some credit, a few gave him very little. These were answers framed and shaped by the debate over patents which was under way as Armstrong was building his gun.

VI

Patent laws clearly affect the pace and direction of technological innovation, as well as the degree of commercial competition in the marketplace for new technologies. But patent laws are also important in determining who receives the credit and the ensuing reputation for an innovation. Reputation is of immense importance in two ways: firstly in terms of personal recognition by one's engineering peers, and secondly to establish a brand name for the commercial production of the new technology. To be first and best greatly eases the struggle for business. But patent rights are socially and politically constructed because the definition of what constitutes a new technology or new technique involves a decision over which the evolving cultural context exerts much influence. Patents are temporary grants by the state for the exclusive use of a technology or activity for commercial profit. They reflect a belief that such protection is a stimulus to inventive activity. That of course is debatable and hence patents are as much political–legal as they are technological. To achieve a reputation the engineer must convince others that his innovation is unique. Armstrong defended his gun for the sake of his engineering reputation; he rejected British patent law as the best means to promote innovation. His response to patent rights demonstrates a sharp entrepreneurial mind.

Like cannons, British patent laws remained unchanged for two hundred years. They were established in 1623 by the Statute of Monopolies, and became the subject of heated debate in the nineteenth century.[61] They were not equipped to deal with the many inventions and developments associated with the Industrial Revolution, and agitation to reform them began in the 1820s. As with the trade laws, there were those who wanted to abolish the patent laws, those who wanted to keep them as they were and those who wanted to reform them. Protectionists (those who wanted to keep the laws unchanged) spoke about property rights, just rewards and individual incentive as the engine of technological progress and social development. Abolitionists considered invention a social phenomenon, part of 'the great mental whole of society', rather than the product of a single individual.[62] Protectionists argued that patent rights protected the individual from having his invention expropriated by society; abolitionists wanted to end the expropriation of the whole of society which 'owned' all the knowledge used by inventors. To the abolitionists the head-start gained by an inventor over all others was reward enough, although they advocated in addition that one-time bonus rewards be paid to inventors by the government and professional associations.

The controversy grew into a widespread public debate which reached its height between 1850 and 1875. Government inquiries were undertaken in 1851–2, 1862–5 and 1869–72. The matter was settled in 1872 by legislation which reformed the system along the lines advocated by John Stuart Mill (and before him Jeremy Bentham and Adam Smith). Patent protection was reduced to seven years and was withdrawn from patents not worked for two years. Abolitionists like Lord Stanley (whose protectionist views changed while he was chairman of the Royal Commission on patents in 1862–5) and *The Economist* were defeated. Patent law continued to be a topic of debate and the courts were kept busy with patent infringement cases, but after 1875 professional organizations and economists turned their attention to protectionism and free trade in the world economy.[63]

Armstrong was an abolitionist. He and Brunel had hardly begun work on their new gun in 1854 when they discovered that James Longridge had taken out a patent on a design very close to the one Brunel had developed whereby the barrel was strengthened by winding wire under high tension tightly around a wrought-iron tube. As in the gun Armstrong eventually made, this technique created centripetal forces on the barrel from the wire which had been stretched but prevented from returning to its original length by the outer wall of the barrel. When he learned of Longridge's patent, Brunel wrote to Armstrong in 1855 to say that he could not pursue their experiments. The principle behind their 'coil' system was sound, he said, 'and but for these patents' they could have made an excellent gun. Despondent, Brunel left the project, convinced that patents destroyed competition and technological progress.[64]

Armstrong pressed ahead by himself. However important men like Rendel and Brunel had been in getting him initially involved in making guns, Armstrong's own determination carried him forward. He joined the movement to abolish the patent laws which, like the Corn Laws, he claimed, constituted class legislation by which a group of non-productive people lived off the labour of others and, at the same time, retarded technological progress and thereby injured the public good:

> That dauntless spirit which in matters of commerce has led this country to cast off the trammels of protection [he said] has resulted in augmented prosperity to the nation, showing the injurious tendencies of class legislation when opposed to general freedom of action. Would that the same bold and enlightened policy were extended, in some degree at least, to matters of invention.[65]

The true inventors, he said, were those who, 'by their labour and intellect, give available reality to ideas' and breathe life into the 'crude ideas which pen-and-ink inventors can produce in unlimited numbers, and which are almost worthless until reduced to practice'.[66] Ideas should be common property, he said, and he called for free trade in 'primary ideas' and 'fundamental conceptions' in which 'the merit of invention seldom lies'. Pen-and-ink charlatans abused the patent laws, while the true inventors, who produced real products for the benefit of society, were left to 'struggle with difficulties, unknown to the mere theorist, and often requiring years of labour, blended with disappointment, for their removal' and were rewarded with only 'tormenting litigation about priority of ideas'. Armstrong wanted this reversed. Authentic inventors should receive patent protection, although it should be 'sparingly awarded'. Finally, he called for the patent laws to be replaced by a system of state grants for those who produced working inventions which others could improve upon without fear of lawsuits.[67]

Among the abolitionists was Sir Nathaniel Barnaby, the Naval Constructor during the early 1860s. At the inaugural meeting of the Royal Institute of Naval Architects he presented the abolitionist case. Many inventors had pleaded or demanded that the Admiralty finance the development of their ideas (or in some cases mere suggestions), but at the same time maintained the patent rights or claimed compensation if the idea happened to prove practical. The 'inventor' thus avoided all risk and expense on behalf of his idea:

> Naval architecture has little need for such inventors as these [Barnaby said]. What it needs is men like Stephenson and Watt, Armstrong and Whitworth, who will patiently work out with their own hands that which their brains have conceived; meeting difficulties at every turn – at every turn devising the means of overcoming them. Such inventors will not lose their reward.[68]

Engineering societies were divided over patents and usually advocated compromise. The Institution of Mechanical Engineers and the Society of Engineers advocated reform, not elimination, of the patent laws.[69] Armstrong's personal experience and his particular engineering abilities may account for his position. One can see his frustration and anger as he worked day and night on his gun, overcoming innumerable difficulties, testing, retesting and slowly bringing an idea of a gun to life, only to be confronted in the end by laws which deprived him of his reward. Years afterwards he recited Brunel's withdrawal in 1855 (without naming Longridge) and expressed sadness that 'people quietly submit rather than incur a troublesome and expensive litigation'. He recalled an experience in 1849, when his crane company won a contract to supply the Great Northern Railway Company with wrought iron wheels welded into one solid piece. After the wheels had been delivered a claimant appeared with a 14-year-old patent on Armstrong's wheel design.[70] Thus pen-and-ink inventors, an old drawing in one hand, the patent laws in the other, waited, for years if need be, to ambush unsuspecting manufacturers.[71]

The armaments community was international and Armstrong was often accused of stealing designs from other nations. A French company claimed that Armstrong's new cannon was their invention and had been registered in the English as well as the French patent office in 1849. Armstrong, the French firm said, had done nothing more than copy its English patent. They readily admitted that Armstrong had a legitimate claim to 'the more or less ingenious arrangements' he added, but they protested in view of 'the noise about the wonderful effects' of Armstrong's gun.[72] Others were less accommodating. *Scientific American* exclaimed: 'Not content with appropriating the ideas of others right and left, he would have larceny legalized by the world.'[73]

Many accused Armstrong of using their designs, but one inventor, Theophilus Blakely, stands out. Blakely, a retired captain of the Royal Artillery, had designed a built-up gun and had taken it to the Ordnance Committee in July 1854 (that is, five months before Armstrong met the Duke of Newcastle). But the government was not interested in artillery at that moment and his proposal was rejected. As we saw earlier, the determined Blakely set off for the Crimean battlefields in search of support from Lord Lyons. This did him no good and so he took out a patent in 1855.[74] Later, Blakely had been granted a few hundred pounds by the government to work on his gun and, when Armstrong's built-up gun won recognition in 1858, Blakely claimed it was his design. He began a long, hard, losing campaign to gain credit for the built-up idea used by Armstrong.[75] Whenever Blakely was present at one of Armstrong's speeches on patents, there was sure to follow an 'animated and lengthy discussion'.[76]

Blakely's struggle was futile. The government was firmly committed to Armstrong and had no intention of allowing Blakely or the patent laws to interfere. Blakely was not allowed to compete against Armstrong and Whitworth and, in December 1863, frustrated and bitter, he gave up on England:

> I have been well-treated abroad, ill-treated here; [he cried] and in future any skill I may possess shall be at the service of those who have acted fairly by me, in preference to this Government, *which has forfeited all claim to my consideration* (original emphasis).[77]

This was not an idle protest. Blakely sold many guns abroad: to both sides of the American Civil War, to Russia, Sweden, Spain, Portugal, Italy and in South America.[78] His gun performed as well as the Armstrong. According to *The Engineer*:

> The Defeat of the Spanish fleet at Callao is undoubtedly a triumph for the Blakely and Armstrong guns (the first more especially), used by the victorious Peruvians. ... It will be Captain Blakely's own fault if he does not make political, or, rather, commercial capital out of the affair.[79]

Success in battle is the ultimate test for weapons, but it came too late for Blakely. The large orders from America ended with the Civil War and, in the 1860s, Armstrong, Whitworth and Krupp entered the world marketplace and small entrepreneurs like Blakely were swept aside. In 1868, at the age of 41, he died in Peru, 'insolvent, and unrewarded by the Nation'.[80]

The government also played the patent game. The accusations made in the 1850s that the War Office stole designs from private inventors were confirmed in the 1860s. After rejecting the designs of private inventors, the War Office and Admiralty coolly used them to build their own mix-and-match versions.[81] Parliamentary petitions outlined how the ideas of different inventors had been blended at the Royal Gun Factories to obtain the cheapest possible gun and, at the same time, avoid paying remuneration to the inventors.[82] The navy did the same. It was doubtful, said Admiral Elliott, that the Surveyor's Office had ever acknowledged to an inventor that his idea was any good.[83] Guns built in the Royal Gun Factory were of poor quality because the Ordnance Select Committee was too occupied with 'inventing and dressing up the ideas of other men as its own, to attend to such trifling matters as the endurance of guns'.[84] This plunder was 'an elaborate art worked out with much skill and ingenuity'.[85] The First Lord of the Admiralty, the Duke of Somerset, readily admitted that he had 'invented' a gun in this manner. He had, he said, merely done 'what many other inventors had done, to take advantage of other people's ideas'.[86] Like Armstrong, the Admiralty felt 'ideas' could be freely adapted and saw nothing wrong with the fact that the navy was both a competitor and the judge.[87] And, as Lord Aberdeen had reminded one MP who complained about this matter in 1854, Parliament could not be sued.[88] There is no evidence that anyone tried to sue the War Office or the Admiralty. The practice continued into the twentieth century.[89] The 1883 Patents Act repealed the right of the

Crown not to pay royalties, but disputes were settled by arbitration at the Treasury Department and the government retained the power to have an innovation assigned to it and kept secret.[90] A few inventors managed to collect either a lump-sum[91] or royalties.[92] Others supplied the government from their own factories, but they had to sign solid contracts and remain vigilant, as the government often tried to manufacture the item without paying royalties.[93]

Armstrong was also caught in this contest. His guns were of the highest quality and consequently expensive, but after Palmerston's death, government policy was, in Cardwell's words, 'to secure the highest ultimate efficiency at the least expense'.[94] However, the Admiralty and the War Office usually sacrificed the former to the latter. Sir Ashley Cooper Key, captain of the training ship *Excellent*, assimilated Armstrong's inventions with those of others.[95] Samuel Goddard, a Birmingham gun maker who had lost out to Armstrong in 1858, was sent to Woolwich by the War Office in 1863 to take from the Armstrong design whatever would improve his own cheaper model.[96] John Anderson, whom we shall meet again, modified Armstrong's rifling method for muzzle-loaders. Armstrong, finally stung, protested. It was, he said, 'unjust to him, the inventor' of shunt rifling, and called the incident 'one of those hasty modifications of an approved invention which are … unfair to inventors, as attaching to their works the names of persons who merely alter without improving'.[97]

But the Government was determined to have cheap guns. Colonel Edward Boxer, Superintendent of the Royal Laboratory at Woolwich, developed fuses for all Armstrong guns which were half the price.[98] Meanwhile, the Assistant Superintendent of the Royal Gun Factory, a Mr Fraser, modified Armstrong's coil construction and built the cheaper Fraser, or Woolwich gun, which Cardwell hailed as the best weapon available.[99] Unfortunately, a Fraser gun burst on board the *Thunderer* in 1878, killing a number of sailors. Armstrong took the opportunity to remind the government that, 'in the choice of a system, what is of primary importance is to attend not to theoretical considerations but practical experience'.[100] Although he admitted the Fraser guns had shown 'extraordinary endurance', he reminded the government that the cheaper Fraser pattern sacrificed the best technology to cost. It was not until the 1880s, when the supremacy of the British Navy was questioned, that the balance tilted the other way.

The reluctance of British governments to pay the price for new guns, while particularly strong in the two decades after Palmerston's death, was evident from the beginning of the armaments revolution in the 1850s. In December 1858, Lord Derby appealed to his Secretary of State for War not to buy too many Armstrong guns, given 'the present infancy of the invention and doubtless time and experience [would] both suggest improvements and produce economy in manufacture'.[101] This of course was sound advice and Armstrong himself advocated that the War Office leave open the prices of his guns as they were falling rapidly. Amidst the excitement over Armstrong's guns in 1858, the government, eager to contain costs, conducted its own 'careful and extensive experiments' to strengthen the old cast-iron guns or to see if 'any better, speedier, or cheaper system of constructing rifled guns existed than that proposed by Sir W. Armstrong'.[102] When they failed, the government was forced to come to terms with Armstrong.

VII

Armstrong invented his gun within the particular historical context of 1850s Britain. The Crimean War provided the impetus to design a gun with particular features; the invention, that is Armstrong's design idea which he believed would break a technological bottleneck, was developed with help and advice from other engineers who, like Armstrong, sought unity and elegance as well as functional efficiency in their designs. Convincing state officials and politicians that his gun was both his own invention and better than its competitors was not a straightforward activity. Certainly, the participants at the time treated technological change as a matter of vigorous argument and very much a social–political process as well as a technological one. The design of Armstrong's first gun reflects many external forces as well as his individual genius. That he survived the experience says as much about his social and political sense as it does about his technological insight. The story so far has looked at Armstrong's gun as an artillery piece. But the Crimean War also had a naval element which began a naval revolution which presented Britain with a fundamental strategic crisis. Armstrong would play a decisive role in the outcome of that crisis, which changed the course of British defence policy for 20 years.

Notes

1. James Rendel was the father of Stuart Rendel; the latter would play an important role in Armstrong's companies.
2. William Armstrong, 'Report on the Construction of Wrought Iron Field Guns' (14 July 1855), W.O. 33/11; Stuart Rendel, *The Personal Papers of Lord Rendel*, ed. F.E. Hamer (London: Ernest Benn, 1931), pp. 269–70.
3. Britain sought to destroy Russian naval power in the Black Sea which threatened free trade on the Danube, Turkish control of access to the

Mediterranean and ultimately British naval supremacy in the eastern Mediterranean. The stakes were very high.

4. *The Times*, 7 Nov. 1854, p. 6; 4 Dec., pp. 6–8 (a lengthy analysis); 28 Dec., p. 9; General Simpson to Lord Panmure, in Sir George B. Douglas and Sir George Dalhousie (eds), *The Panmure Papers: Selections from the Correspondence of Fox Maule, 2nd Baron Panmure* (London: Hodder and Stoughton, 1908), vol. 1, p. 170; *Report of the Artillery Committee on Ordnance Employed at Sebastopol* (17 Nov. 1855), W.O. 32/7555. The Russian naval guns, taken into Sebastopol fort from their ships sunk in the harbour, had a range and power advantage over British and French field guns.
5. Alexander W. Kinglake, *The Invasion of the Crimea* (New York: Harper and Row, 1875), vol. 3, pp. 255–6. This 8-volume work provides details of the military operations. For a summary of operations, see Sir E. Hamley, *The War in the Crimea* (London: Seeley, 1891).
6. Charles Ffoulkes, *Arms & Armament: An Historical Survey of the Weapons of the British Army* (London: George G. Harrap and Company, 1945), pp. 93–5.
7. Geoffrey Parker, *The Military Revolution: Military Innovation and the Rise of the West, 1500–1800* (Cambridge: Cambridge University Press, 1988), pp. 10–14; Charles Oman, *A History of the Art of War in the Sixteenth Century* (Elstree, Hertfordshire: Greenhill Books, 1987), p. 28.
8. Parker, *The Military Revolution*, p. 32. Parker has Ireland in mind but he also concludes (p. 33) that the English, who applied enormous firepower in Ireland in the sixteenth century, 'still lagged far behind continental practice at home'.
9. Oman, *Art of War*, p. 357; Parker, *The Military Revolution*, pp. 26–8.
10. Theodore Ropp, *War in the Modern World*, 2nd edn (New York: Collier Books, 1962), pp. 24–5; Parker, *The Military Revolution*, pp. 9–10.
11. Parker, *The Military Revolution*, p. 23.
12. Ibid., p. 151; see also Gunther E. Rothenberg, *The Art of War in the Age of Napoleon* (London: B.T. Batsford Ltd., 1977), p. 26.
13. Rothenberg, *The Art of War in the Age of Napoleon*, pp. 74–5, 143–5, 156.
14. Parker, *The Military Revolution*, p. 33.
15. Hew Strachan, *European Armies and the Conduct of War* (London: George Allen & Unwin, 1983), pp. 111–13.
16. F.S. Stoney and C. Jones, *A Text-Book of the Construction and Manufacture of the Rifled Ordnance in the British Service*, 2nd edn (London: HMSO, 1872), p. 2.
17. Ropp, *War in the Modern World*, p. 145; Gerhard Ritter, *The Sword and the Sceptre: The Problem of Militarism in Germany*, vol. II: *The European Powers and the Wilhelminian Empire, 1890–1914*, trans. Heinz Norden (Coral Gables, Florida: University of Miami Press), pp. 33–4.
18. Peter Cadogan, *Early Radical Newcastle* (Consett: Sagittarius Press, 1975).
19. As high, for example, as his brother-in-law, who became a baron of the Exchequer. Armstrong probably met Sir Robert Peel in 1847: see Peel Papers, Add. Ms. 40599, f. 219.
20. Peter McKenzie, *W.G. Armstrong: The Life and Times of Sir William George Armstrong* (Morpeth: Longhirst, 1983); *Encyclopaedia Britannica* (1910).

21. *D.N.B.* vol. 22 (Supp.); David Dougan, *The Great Gunmaker: The Story of Lord Armstrong* (Newcastle: Frank Graham, 1970).
22. Armstrong continued to write professional papers in his retirement, working until he died at the age of 90 in 1900. He published *Electric Movement in Air and Water* in 1897.
23. Armstrong's armaments empire began with hydro-electric machinery. After the First World War, when demand for armaments evaporated, the company failed to make the transition back to civilian engineering with its purchase of a hydro-electric project in Newfoundland.
24. *A Visit to Egypt* (1873).
25. William Armstrong, *Report on the Construction of Wrought Iron Field Guns* (14 July 1855), W.O. 33/11; Stuart Rendel, *The Personal Papers of Lord Rendel*, ed. F.E. Hamer (London: Ernest Benn, 1931), pp. 269–70. Armstrong's version accentuates his own part while James Rendel's son, Stuart, writing many years later, emphasises the role of his father. Armstrong, for example, suggests that he saw the Duke alone while Stuart Rendel says that his father took the plans to the Duke on behalf of Armstrong.
26. See two essays by Clive Trebilcock: 'Science, Technology and the Armaments Industry in the UK and Europe, with Special Reference to the Period 1880–1914', *Journal of European Economic History* 22 (1993), pp. 565–80; 'Spin-Off in British Economic History: Armaments and Industry, 1760–1914', *Economic History Review* 2nd series, 22 (1969), pp. 474–90. On technological momentum, see Thomas Hughes, 'Technological Momentum', in Smith and Marx, *Does Technology Drive History?*, pp. 101–13.
27. Whereas the Crimean War turned British and French artillery design to rifling light field guns for greater offensive manoeuvrability, the American Civil War turned the attention of American military engineers towards strengthening coastal defences against naval bombardment and siege. This meant focusing on increasing the size and muzzle velocity of smooth-bored guns to be fixed in coastal fortifications. Another significant difference is that the most innovative European artillery engineers were civilian industrialists such as Armstrong and Whitworth in England, Krupp in Germany and Schneider in France, while the leading American artillery innovators were all soldiers, in particular Major T.J. Rodman and Major Parrott.
28. *Hansard*, 11 May 1855, col. 402; 27 July, cols 1458–9.
29. 'Ordnance Select Committee and Inventors', *The Times*, 12 May 1855, p. 10.
30. *The Times*, 19 April 1855.
31. 'Improved Artillery', *The Times*, 2 Aug. 1855, p. 6.
32. Theophilus Blakely, *A Letter from Captain Blakely, H. P., Royal Artillery, to the Secretary of State for War Claiming the Original Invention of an Indispensable Feature of The Armstrong Gun* (London: J. Ridgway, 1859).
33. *The Times*, 2 April 1855, 19 April 1855.
34. 'Approved But Not Adopted', *The Times*, 12 May 1855, p. 10; 'A Gunner', 28 Dec. 1854, p. 9; Bashley Britten, letter, 2 Aug. 1855, p. 6.
35. Letters from George Hale and George Mills to *The Times*, 14 Nov. 1854, p. 10 and 9 May 1855, p. 7. *The Times* became a clearing house for communications between individual inventors, including, as we shall see,

an important one between Armstrong and Sir Joseph Whitworth. The paper closely followed developments and debates over design in armaments technology.

36. Celia Brunel Noble, *The Brunels: Father and Son* (London: Cobden-Sanderson, 1938), pp. 207–9.
37. *Correspondence Respecting the Reorganization of the Ordnance Select Committee*, W.O. /33/1, No. 1, (8 Jan. 1855), p. 101.
38. Douglas and Dalhousie (eds), *The Panmure Papers*, vol. 1, pp. 177–8, 197, 203, 210.
39. *Correspondence on Reorganization* (1855), W.O. 33/1, No. 2, pp. 103, 105, 109.
40. Members of the Committee of Military Officers and Newcastle's Ordnance Committee which replaced it are listed (by position) in the 'Royal Commission into the System Under Which Patterns of Warlike Stores are Adopted and the Stores Obtained for Her Majesty's Service' (Warlike Stores Commission), *Parl. Papers* XV (1887), Q. 941. For the names of a few individuals on the new committee, see *Hansard*, 11 May 1855, col. 402.
41. Ross to Roebuck Committee, Qs. 18007–09.
42. See Ross evidence in 'Reports from the Select Committee on the Army Before Sebastopol' (Roebuck Committee), *Parl. Papers* IX (1854–5), Qs. 18035–38.
43. *The Times*, 19 April 1855; *Hansard*, 2 Aug. 1855, cols 1656–8.
44. W.O. 33/1, 6 Feb. 1855.
45. See the 'Memorandum For Inventors' issued by the War Office in 1859, *Parl. Papers* XXXII (1863), pp. 238–9.
46. 'Warlike Stores Commission', Q. 942; O.F.G. Hogg, *The Royal Arsenal: Its Background, Origin, and Subsequent History* (London: Oxford University Press, 1963), app. XVI, vol. II, pp. 1435–6.
47. I. Brunel, *The Life of Isambard Kingdom Brunel* (London: David and Charles, 1971), pp. 454–61.
48. *The Times*, 7 Jan. 1853, p. 6.
49. Arthur Hamilton-Gordon, Lord Stanmore, *Sidney Herbert of Lea: A Memoir* (London: Murray, 1906), vol. 1, pp. 178–9. Herbert's opinion would be very different during the invasion panic of 1859.
50. Bashley Britten received £734, Haddan £435, Lawrence £266, Lyonal Thomas £398, Montigny £266, Lieutenant Engstrom £500 and Goddard £652.
51. 'Return of the Amount of Public Money advanced since 1852 to Private Persons for the purpose of enabling them to make Experiments for the purpose of Improving WEAPONS of WAR', *Parl. Papers* XLI (1860), p. 657. Nasmyth announced that the government had 'empowered' him to develop his wrought iron gun in *The Times*, 7 Dec. 1854. See also *Hansard*, 29 Jan. 1855, col. 1065.
52. Speech to the British Association, *The Times*, 7 Sept. 1861, p. 7.
53. Brunel, *Life of I.K. Brunel*, pp. 84–6, 452–3. There is no evidence that any legal or business agreement existed among the three men.
54. Armstrong's hydraulic crane, which had produced his first fortune 20 years earlier, was based on the designs of another brilliant engineer, Joseph Bramah. See Ian McNeil, *Joseph Bramah: A Century of Invention, 1749–1851* (Newton Abbot: David and Charles, 1968), pp. 128, 156.

55. As we shall see, Armstrong developed a rifled muzzle-loader later.
56. Ian V. Hogg, *The Illustrated History of Ammunition* (Secaucus, NJ: Quarto, 1985), pp. 29–31.
57. In the late 1860s and early 1870s, after improvements in steel quality were made, a steel inner tube was used. Krupp made cast steel guns and refused to give them up despite the fact that many of his early guns burst when fired more than a few times. See William Manchester, *The Arms of Krupp 1587–1968* (New York: Bantam, 1970).
58. When a cast iron or steel gun is fired, each concentric layer of uncompressed metal absorbs a strain on the barrel inversely proportionate to the square of its distance from the axis of the bore. The limit is soon reached beyond which additional layers or thickness contribute nothing to the strength of the gun and, indeed, act as dead weight. Armstrong claimed that, in his built-up gun, the outer layers of the walls received as much of the stress from the explosion as did the inner layers. See Armstrong, 'Report on the Construction of Wrought Iron Field Guns'.
59. Armstrong described his gun in his 'Report on the Construction of Wrought Iron Field Guns', W.O. 33/11 and in *The Times*, 25 Nov. 1861, pp. 12, 17. See also Dougan, *Great Gunmaker*, pp. 57–61, and Frederick Robertson, *The Evolution of Naval Armament* (London: Constable, 1921), pp. 200–201; Stoney and Jones, *Text-Book of Construction*, ch. 2; *Encyclopaedia Britannica*, 9th edn (1890), vol. 11, p. 288.
60. *Confidential Report*, W.O. 33/9, p. 813.
61. Victor Batzel, 'Legal Monopoly in Liberal England: The Patent Controversy in the Mid-Nineteenth Century', *Business History* 22 (1980); Fritz Machlup and Edith Penrose, 'The Patent Controversy in the Nineteenth Century', *Journal of Economic History* (May, 1950), pp. 1–29.
62. *The Economist*, 28 Dec. 1850, p. 1434, quoted by Machlup and Penrose, 'Patent Controversy', p. 15.
63. Armstrong's company was sued from time to time for patent infringement. See, for example, *Reports of Patent, Design, Trade Mark, and Other Cases*, vol. 20 (1903), pp. 523–4; vol. 23 (1906), pp. 666–719; vol. 24 (1907), pp. 593–640.
64. McNeil, *Joseph Bramah*, p. 454. This was probably not the only reason he left. Brunel was becoming absorbed in his *Great Eastern* project about this time.
65. Armstrong, speech to members of the Mechanical Engineers Society, *The Times*, 2 Aug. 1861, p. 7.
66. Armstrong, speech to the Mechanical Engineers, *The Times*, 7 Sept. 1861, p. 7.
67. Ibid.
68. Nathaniel Barnaby, 'On Mechanical Invention in its Relation to the Improvement of Naval Architecture', Royal Institute of Naval Architects, *Transactions* 1 (1860), p. 158.
69. Batzel, 'Legal Monopoly in Liberal England', *Business History* 22 (1980), p. 193.
70. Ibid.
71. The abolitionists cited numerous examples of crafty patent entrepreneurs who obtained patents for vague designs from the government patent commissioners (who collected a large fee). See Barnaby, 'On Mechanical Invention', pp. 156–7. There were many charlatans no doubt, but James

Longridge was not one of them. Brunel considered his wire-wound gun a substantial step forward that deserved respect. The wire-wound method provided a greater uniformity of stress than hoops or coils and was used again in the 1890s. See W. Hovgaard, *Modern History of Warships* (London: E. and F.N. Spon, 1920; reprint edn, 1971), p. 391. H. Palin, *Memoir of the Late Lord Armstrong* (London: Gurney, 1901), says Armstrong used the Longridge patent when it expired.

72. *The Times*, 7 April 1859, p. 10.
73. Cited by F.E. Comparato, *Age of Great Guns* (Harrisburg, Pa.: Stackpole, 1865), p. 317.
74. *The Times*, 7 Dec. 1863, p. 10.
75. Theophilus Blakely, *A Letter from Captain Blakely*.
76. *The Times*, 7 Sept. 1861, p. 7.
77. Quoted in George Rendel, 'Guns and Plates', *Quarterly Review* 115 (1864), p. 139.
78. James Hazlett, Edwin Olmstead and M. Parks, *Field Artillery Weapons of the Civil War* (Newark: University of Delaware Press, 1983), p. 197; Warren Ripley, *Artillery and Ammunition of the Civil War* (New York: Van Nostrand Reinhold Company, 1970); Mrs T. Blakeley, *Blakeley and Armstrong Guns: A Statement* (London: n.p., 1898); Theophilus Blakely, Letter to the Secretary of State for War (1859); *Hansard*, 2 March 1865, col. 972.
79. *The Engineer*, 15 June 1868, p. 437.
80. *The Memorial of MPs to Captain T.A. Blakeley* (London, n.p., 1901), reproduced in Mrs. Blakeley, *Case for 'Justice'* (1902), p. 15. Mrs Blakeley's plea is a sad footnote to her husband's struggle to enter big-time armaments business. Thirty years after his death, Blakeley's widow took the occasion of Lord Armstrong's death to renew her campaign to have the government, the Sir W.G. Armstrong Company and *The Times* recognise her husband's contribution to the development of the first cannons and provide a pension for his old and poor widow. As the above reference indicates, she was not without political support. See also Mrs T. Blakely, *Blakely and Armstrong Guns*. In 1867, a sympathiser of Charles Lancaster reported that 'Mr Lancaster was dying "worn out and used up", as he said, from fruitless efforts to procure proper acknowledgment of the worth of his gun'; *Hansard*, 6 June 1867, col. 1706.
81. *Engineer*, 18 May 1866, p. 362.
82. *Hansard*, 16 July 1868, cols 1250–54.
83. Christopher Lloyd, *The Nation and the Navy* (London: Cresset Press, 1954), p. 215.
84. *Engineer*, April, 1866, p. 248.
85. *Engineer*, 18 May 1866, p. 362.
86. *Hansard*, 18 April 1864, col. 1180.
87. James P. Baxter, *The Introduction of the Ironclad Warship* (Cambridge, Mass.: Harvard University Press, 1933), p. 130, fn.1; *Hansard*, 11 May 1868, cols 26–32.
88. *Hansard*, 10 July 1854, cols 1420–26.
89. Clive Treblicock, 'A "Special Relationship": Government, Rearmament, and the Cordite Firms', *Economic History Review* 19 (1966), p. 375; Jan T. Sumida, 'British Capital Ship Design and Fire Control in the Dreadnought Era', *Journal of Modern History* 51 (June 1979), pp. 205–30.

90. A copy of the Act, along with discussion by correspondence within the War Office, is in W.O. 32/6206. See also *Hansard*, 16 Feb. and 16 April 1883.
91. *Hansard*, 13 July 1868, cols 1106–7.
92. Stanley Sandler, *The Emergence of the Modern Capital Ship* (Newark, Delaware: University of Delaware Press, 1979), p. 292, fn. 7. See also the negotiations by which, in 1882, the government acquired the rights to the Brennan torpedoes in W.O. 32/8585.
93. Charles Lancaster's long struggle over payments and patent rights are a good example. See the 'Abstract of Proceedings having reference to Mr. Lancaster's Inventions and Claims', W.O. 33/7.
94. Ibid., col. 1134.
95. Sandler, *Emergence*, pp. 103, 107, argues that Key resisted the new gun technology, but it is more accurate to say that he wanted changes made in the Armstrong system. He wanted to maintain the old tactic of improvement by attaching bits and pieces from other inventions.
96. *Hansard*, 16 July 1868, cols 1251–2.
97. Adm. 1/5905.
98. 'Papers Relating to the Dismissal of Colonel Boxer', *Parl. Papers* XLII (1870), pp. 480–81.
99. *Hansard*, 11 March 1869, col. 1133.
100. Stuart Rendel, *Memo of Verbal Observations in Connection with the Bursting of the 38 ton Gun on Board the 'Thunderer'*, T&W, 31/5082.
101. 'Report of the Select Committee on Ordnance' (July 1863), *Parl. Papers* XI (1863), pp. iv–v.
102. Ibid.

CHAPTER TWO

The naval arms revolution

I

Armstrong's artillery piece began a military revolution of great consequence. By breaking through the technological limitations that heavy, brittle cast iron had imposed upon artillery, the Armstrong wrought-iron gun opened up a whole range of possibilities for the imaginative strategist and tactician. The rifleman lost his advantage of range, manoeuvrability and accuracy of fire. The natural barriers which had protected vulnerable peoples were overcome in the imaginations of the new imperialists. The lightweight but powerful Armstrong gun could be easily dismantled and transported by mule up and over Asian mountains and by river boat up river and pathways through the tropical forests of Africa. Armstrong's new gun was first an imaginative idea to overcome engineering problems related to building a wrought-iron gun. But, once built and its power proved, the Armstrong gun stimulated other imaginations about the various possible uses for the weapon. The gap is narrow between Armstrong's gun and the imperialist imagination.

The idea for his gun was a direct result of the Crimean War and the new journalism which reported on that war and Armstrong's own desire to take on new engineering challenges. But, once built, the Armstrong rifled breech-loader stimulated other imaginations to devise new ways to use the weapon. Armstrong set out to remake British military power, but his new gun was quickly seen as also a means to remake British naval power. His first guns were small, but they could in principle be built up into very large guns for use in land fortifications. But Armstrong incorporated into the gun the hydraulic machinery that he had developed for his crane business, and it was this combination that began a naval revolution. While gun design and engineering was the focus of attention until about 1880, it was hydraulic machinery to manipulate the enormous guns that became the main point of the innovator's imagination. And, as we shall see, here too Armstrong's hydraulic technology made Elswick indispensable in the rebuilding of Britain's fleets.

By adapting his artillery gun for use on warships, Armstrong played his part in a long historical process. Guns had always been designed first for armies and then taken on board warships, a change which required warships to be redesigned to accommodate guns. Warships are little more than floating gun platforms (or fight decks or missile launch pads) and naval architecture is determined by gun technology. And yet technology is more fundamental to navies than armies: sailors can neither walk to the battleground nor find safety in the natural environment. Naval forces are highly concentrated; great important battles can involve relatively few but expensive ships. Warships, whether powered by sail or by engines, are complex technological systems that require highly-skilled personnel to operate effectively. The loss of a single battleship considerably weakens a fleet. The naval revolution of the nineteenth century was driven by the realisation that a technological innovation could overcome disadvantages of the existing technology, and rather suddenly. Thus naval 'panics' and naval arms races were a feature of the military revolution of 1854–1914 (just as ICBM panics and races accompanied the rise of nuclear missiles in the second half of the twentieth century).

The nineteenth-century naval revolution, in which Armstrong played a leading part, was the first modern version of an historical phenomenon, dating back to the artillery innovations of the fourteenth and fifteenth centuries which made their way into European navies during the sixteenth and seventeenth centuries, introducing a new era of naval power based on large, heavy sailing ships whose sides rose up from the water like great wooden walls out of which protruded rows of heavy cannons. A galley could carry six or seven cannons but even a small sailing ship could carry 40 cannons and by the seventeenth century large ships carried 100 guns or more. The first 'great ships' were built in the first half of the sixteenth century and they rapidly grew in size over the next hundred years. The *Sovereign of the Seas*, built with Charles I's ship-money, had three decks to hold three rows of heavy guns above the water. Its 1500-ton displacement and great height made rowing impossible. The age of sail had begun.

The introduction of heavy guns altered the nature of naval warfare as battles between coastal-bound galleys, which involved ramming, boarding and hand-to-hand combat, were replaced by thunderous artillery duels between sailing ships on the high seas. The aim of sea battles shifted from that of killing men with sword and musket to the sinking or burning of ships with shot and shell. Combat between men was replaced by combat between ships. Ships were built for firepower and decks were added to carry more guns. The 100 guns of the *Sovereign of the Seas* were deployed for concentrated fire at close range while her thick sides were left to absorb the shot from the enemy.[1] A hierarchy of warships emerged after the sixteenth century: at the pinnacle sat the ships-of-the-battleline, the 'great ships' of the seventeenth to the early nineteenth century. For nearly three centuries war often included massive gun battles between ships separated by less than a hundred yards. These mighty ships were created by the heavy gun and when naval guns outgrew them in the second half of the nineteenth century they were replaced by the modern battleship.

Little new technology was added to naval warfare during the eighteenth century. Past technological innovations were developed, refined and brought to their peak of perfection. The eighteenth century has been aptly named the 'neoclassical age of modern warfare'.[2] By 1700, the fundamental technological innovations of the past had reached their limits, and over the next hundred years strategy, tactics and training caught up with technology.

Ships were divided into six classes, or 'rates', according to the number and size of guns carried. A first-rate three-decker warship carried over 90 guns; a two-decker carried over 60 guns; a fourth-rate ship had between 30 and 60 guns; and so on down to fifth- and sixth-rate sloops and corvettes of 20 guns.[3] Naval guns were also divided into groups and subgroups according to their size and shot. They ranged from cannons and demi-cannons which threw 50-pound and 32-pound shot, respectively, to the more manoeuvrable culverins and demi-culverins which fired 24-pound shot. High-trajectory mortars fired pieces of iron or stone against enemy boarders. In the eighteenth century, these medieval names were replaced and guns were identified by the size of shot they threw: 42-pounders, 32-pounders (the old demi-cannon), 18-pounders (the medieval culverin) and so on down to 5- and 3-pounders. Mortars retained their name.[4]

A ship-of-the-line of 1800 was not much different from one built in 1700: 'there were no secret weapons; there was no problem of obsolescence'.[5] The historic naval battles of the Napoleonic wars of the early nineteenth century were firmly of an era which began with the Spanish Armada of the sixteenth century, developed in the Anglo-Dutch conflicts of the seventeenth century, and refined during the Anglo-French wars of the eighteenth century. The Battle of Trafalgar in 1805 brought this age of naval battles among great sailing ships to its climax. It was the last of its kind.[6]

As Nelson lay dying in the cockpit of the *Victory*, the Industrial Revolution was pouring the foundation for a second and more dramatic naval revolution. It had two main technological components. Firstly, as in the transformation of the fifteenth and sixteenth centuries, the means of propelling ships changed as sail gave way to steam power. Secondly, and unlike the changes in the early modern period, new materials were used. Iron, and later steel, replaced wood in the construction of ships. The role played by the British Navy in spreading the new innovations was also different. It had led the way into the age of sail, but resisted adopting the industrial innovations of the nineteenth century despite the fact that England was the source of most of those innovations and had the industrial capacity to produce them in large quantities. However, when others adopted the new means of exercising naval force, which threatened the security of Britain, the Admiralty not only decisively met those challenges but pushed the technologies behind them to new levels.

II

The steam engine, perhaps the most important individual innovation of the Industrial Revolution, was also one of its first inventions. Thomas Newcomen developed his steam engine in the first decade of the eighteenth century and James Watt built his engine in 1769. Steam pressures increased rapidly and William Symington was able to build a prototype steamboat in 1788, and one capable of towing canal barges in 1802. Five years later, the first commercial success was attained by Robert Fulton, whose paddle steamer *Clermont* was driven up and down the rivers and across the lakes of America by a Boulton & Watt engine. But the enormous deadweight of the machinery, and the inefficiency with which these early steam engines converted heat energy from coal to mechanical power, impeded the speedy development of ocean-going steamships. The massive high beam of the early steam engines raised the centre of gravity of deep-draught ocean ships beyond the level of stability. Nevertheless, inventor-businessmen made steady technological progress and steamships crossed the Atlantic regularly after the famous crossing of the *Sirius* and I.K. Brunel's *Great Western* in 1838.

Ship owners turned from wooden sailing ships to iron steamships for several reasons: wooden hulls needed to be as heavy as their cargoes, whereas iron hulls carried twice their weight in cargo, giving increased carrying capacity from lighter ship weight; joints in wooden hulls gradually loosened with the movement of the ship and especially from the pounding of a screw propeller, whereas iron hulls withstood those vibrations; iron ships could combine size and speed far beyond the capabilities of wood. These technical advantages increased efficiency and lowered costs, the prime aim of ship owners. Iron was thus superior to wood as ship building material and the inner qualities of iron determined the design of ships and the skill required of captains and crews. Here iron proved even more a wondrous material in that it could be shaped and modified to fit a desired design. In other words, unlike wood, iron could be shaped into virtually any form and its chemical composition altered by the refining process to produce a wide range of characteristics for various uses. Unlike wood, iron could be altered to suit almost any design idea of an engineer. Great mechanical engineers like Lowthian Bell and William Armstrong realised that imagination now played a greater role in engineering than ever before. In this new world boundaries between civilian and military and the pursuit of engineering itself – inventing new combinations of material and design – became an end in itself, with business a vehicle by which they could continue as engineers. Later in the century that balance would shift, but never entirely reversed.

The British Admiralty kept a watchful eye on the evolution of the steamship. The technical problems faced by commercial ship builders were compounded in naval ships whose operations required greater efficiency and stability. These problems were of particular concern to the British Navy, whose duties expanded after 1815 to global proportions. Furthermore, the Lords of the Admiralty were persuaded that the spread of steam-driven warships did not favour British strategic interest. Freedom from the constraints of wind and tide was obviously desirable for a navy, but the Admiralty saw no relative advantage if all navies adopted steam power. The British naval supremacy that had been consolidated at Trafalgar rested upon sail power and a change to steam might reduce the relative position of British sea power. With France and America in mind, the Lords of the Admiralty issued their famous memorandum of 1828:

> Their Lordships feel it their bounded duty to discourage to the utmost of their ability the employment of steam vessels, as they consider that the introduction of steam is calculated to strike a fatal blow at the naval supremacy of the Empire.[7]

There were tactical disadvantages to the immediate adoption of steam power by navies. The paddle-wheel propulsion system of the first steamships was vulnerable to enemy fire and, since most paddles were on the sides of the ships, the space available for guns was reduced and their arc of fire restricted. Nevertheless, the Admiralty ordered a number of small paddle-wheel steamships in the 1820s and 1830s. The vulnerable paddles were eliminated by the screw propeller in the 1830s. Screw ships not only placed their propellers safely out of the way but were faster and more powerful than paddle-wheelers, a fact demonstrated by the celebrated race and tug-of-war between the screw ship *Rattler* and the paddler *Alecto*, promoted by I.K. Brunel.[8] Screw propulsion opened the way for steam power to be introduced into larger warships in the 1840s.

The Admiralty proceeded with caution but the French, who saw steam power as the technological 'equaliser' with which they could at last challenge British naval dominance, added steam engines to their frigates as soon as was technologically possible. The first was converted in 1844. The perceived disruption of the balance between British and French naval power incited the first 'invasion panics' in England in 1844, 1846 and 1848. The British added steam to their frigates in 1845–6.

A long transition period began in which steam gradually replaced sail and the great ships of the sailing era underwent their industrial metamorphosis into the battleships of the modern world. The first British battleship built as a steamer, the *Agamemnon*, was completed in 1852, but it was still fully-rigged. The *Devastation*, built in 1873, was the first battleship designed to fight under steam alone. Steam began as an auxiliary source of power and warships remained fully-rigged until the increased size and power of naval ordnance brought about the demise of the broadside ship and gave birth to the turret ship. The latter's guns were situated on deck and required that sails, masts and shrouds be eliminated. The great warships of the late-Victorian and Edwardian navies were propelled by engines alone and the vast marine-engineering industry that had developed during the previous hundred years.

III

The use of iron in warships was also initiated by civilian engineers. Inventor-businessmen of the early Industrial Revolution exercised their technological ingenuity at its limits while the British Admiralty waited until the French navy adopted the new technologies before incorporating them into British warships.

The British ironmaster John Wilkinson launched an iron-hulled canal barge in 1787; an iron steamship first crossed the English Channel in 1822. Shipbuilders realized that steamships strengthened with iron could withstand the extra strains set up by the weight and vibrations of their engines. Steamships required a larger hull capacity than sailing ships to accommodate their engines and coal but wooden ships had reached the limits of their size and strength. The only way forward was to use more iron. There were many problems to overcome associated with the relationships among weight, engine power, fuel capacity and displacement, but by the 1840s the era of iron ocean-going steamships had dawned.

Consistent with the past, early nineteenth-century developments in ordnance raised the question of iron warships. The impetus came in 1822, when General Henri-Joseph Paixhans published a book which argued that explosive shells would eliminate the numerical advantage held by the British navy. Old cast-iron mortar bombs, fired with low velocity and high trajectory, could be redesigned, he said, to fire with high velocity and flat trajectory from long naval guns. The French navy supported the development and testing of this design idea, and two years later Paixhans successfully demonstrated his shells. Diffusion followed and by the 1840s shell guns were part of weaponry of the battleships of all the great powers. Shells were used by the Royal Navy with devastating effect against the Burmese in the 1820s and the Chinese in the 1830s and 1840s. The Paixhans shell stimulated the building of a few small iron warships and tests were conducted on iron plates in France and Britain in the late 1840s and early 1850s,[9] but it was the widely-publicized destruction of the Turkish fleet by Russian shells at Sinope that showed that iron was the only defence against the explosive shell. The exigencies of the Crimean War which followed accelerated the trend towards iron warships. France remained a leader in shell technology for both large and small guns, and in 1895 the French navy became the first to carry high-explosive shells.[10]

Britain and France declared war on Russia in March 1854, and formulated their alliance a few weeks later. Their fleets, which had entered the Black Sea in January, were ordered to attack the Russian naval base and fortress at Sebastopol. The Russian fleet was outnumbered two to one and withdrew into the harbour. Its large naval guns were removed into forts and seven of its ships were scuttled across the entrance to Sebastopol harbour to block the allied fleets. On each side of the entrance stood a Russian fort: the Russians settled in for siege warfare. The allies opened their first bombardment of the war in October. However, shallow water kept the deep-draught two- and three-decker battleships beyond the effective range of their heavy guns. On the other hand, they were within effective range of Russian red-hot shot and the Paixhans explosive shells which rained down upon the great wooden ships, setting several ablaze and forcing a withdrawal.

The French had taken heed of the battle of Sinope and had developed plans for an iron vessel. But French industrial capacity was insufficient to produce enough vessels quickly and thus they showed their designs to their new ally and pressed them to employ British industrial might in the production of vessels protected with armour. The First Lord of the Admiralty, Sir James Graham, ordered his own tests of iron plates in September, but finally agreed to build vessels on the French plan. Britain in turn provided France with her plans for mortar vessels, gunboats and even the standard 68-pounder naval gun.[11]

The vessels were of simple design and were speedily built over the winter of 1854–5. The first was launched in March 1855 and others quickly followed. These flat, ugly vessels did not merit the name of ship and were referred to as floating batteries. But their shallow draught allowed them to enter the coastal waters of the Black Sea. Each had simple rigging and a steam engine which delivered four knots speed. Most importantly, they were protected with four inches of iron plating bolted to 18 inches of timber. Each carried sixteen 68-pounder guns. They cost £50 000 each, about half the price of a wooden frigate. Britain built eight, France five.[12]

The vessels made up in performance what they lacked in beauty. In the summer of 1855 they attacked important targets beyond the reach of the great ships. When Kertch, the fort which protected the entrance to the Azor Sea, fell to the allies in 1855, the floating batteries roamed the coasts and rivers at will, destroyed supply routes and depots and sealed the fate of Sebastopol. The ultimate test of their armour came in October 1855, at the attack on Fort Kinburn which commanded the estuary of the Bug and Dnieper rivers. The British floating batteries arrived too late to participate in the attack (the fort fell in under six hours), but the French vessels proved conclusively that four inches of wrought iron on 18 inches of timber was impervious to shot and shell. Russian projectiles fired from under 1000 yards bounced away or broke apart on impact.

The fall of Kinburn spelled the end for the Russians in the Crimea and it seemed apparent that Kronstadt would fall when the allies transferred their main effort to the Baltic in 1856. The Tsar decided peace was in order and in 1856 agreed to terms.

The Crimean War was an historical turning point in many ways. Supply lines and medical organisation took on new importance. It was the first war in which newspapers played an important role. When William Armstrong read newspaper reports of the ineffectiveness of British artillery, the first step was taken towards the invention of the Armstrong gun. Old weapons were found inadequate and new ones subjected to trial by battle. Technology and industrial capacity were shown to be crucial ingredients of victory. Although the most famous incident of the war centred on the individual bravery of the men of the Light Brigade, the fundamental significance of the Crimean War for the future was that the individual inventiveness of engineers which it brought forth could not be easily recalled.

IV

After the war the British Admiralty laid up its ships and floating batteries. The French, however, did not rest. In their eyes three-decker wooden battleships had been rendered obsolete and they stopped building them. Floating batteries were only a first step towards an ocean-going iron navy. Tests on iron plates continued under the personal eye of Napoleon III, and in 1857 plans were drawn up in secret for an ironclad frigate. In the meantime the French pressed ahead with the installation of steam engines in frigates.

The Surveyor of the British navy, Sir Baldwin Walker, watched events in France closely. He had built the British floating batteries and had spent enough time in France to know their interest in developing larger iron ships. He agreed with the Admiralty that technological changes in warships were not in the interests of Britain, but he realised the importance of Kinburn and prepared a design for a British ironclad frigate.

The British Admiralty and its First Lord, Sir Charles Wood, were not eager to encourage expensive technological changes in the world's navies. They did not wish to advertise the broader implications of the floating batteries at Kinburn.[13] Like steam in the 1820s, ironclad frigates in the 1850s would place all the great powers on the same level and erode the existing British advantage. Conversion programmes were also expensive. An ironclad frigate cost £80 000 more than a wooden battleship.[14] Furthermore, technological innovation produced obsolescence and resulted in a spiral of higher costs and taxes. Britain, with the largest navy, would pay the most to keep up. The Admiralty therefore set aside Walker's proposal for an ironclad.[15]

The issue of conversion to steam power had similar problems. The Admiralty had ordered a few steam sloops in the 1820s and, when it became technologically possible to convert frigates in the 1840s, it responded to French conversions with its own. But conversion programmes did not find support among British taxpayers until they perceived an immediate danger. The contest between the *Rattler* and the *Alecto* had been organised to generate public and parliamentary support for the expensive installations of screw propellers,[16] but it required a few invasion panics to attract undivided attention. A massive conversion programme in the post-Crimean years would cost more than the British taxpayer was willing to pay without a good reason.

Walker was convinced he had a good reason. He was a firm believer in steam power and he considered it more important than ironclads, which had not yet actually appeared. Walker was not a ship builder. When he was informed in 1848 of his appointment to the position of Surveyor he thought there was some mistake. But to Lord Auckland, the First Lord at the time, Walker's lack of technical training was secondary to his 'judgement' and 'high character' and he did not possess the 'prejudice' of 'theoretical' knowledge which kept out new ideas. What this meant was that Walker would oversee the conversion of frigates to screw power, which his predecessor, Sir William Symonds, had refused to do.[17]

Ten years later, Walker again counted steamships and became convinced that France would soon have more than Britain. In 1858, Britain and France had 30 steam frigates each, but Walker calculated that by 1860 France would have four more than Britain (40 to 36). He proposed the immediate conversion of six frigates and called for a future fleet of between 60 and 80 steamships.[18]

In March 1858, the Palmerston government resigned, a victim of its Conspiracy to Murder Bill, and the Conservative Party led by Lord Derby formed a minority government. A Tory government was not necessarily a good thing for Walker. He was a Whig appointee and had antagonised Derby in 1852 by successfully resisting attempts of the first Derby Admiralty to practise patronage in the dockyards.[19] However, the naval situation was rapidly changing. The rumours which circulated in March 1858 that the French were about to build an ironclad frigate were confirmed in May. Walker prepared a report for the new Tory Board of Admiralty and the First Lord, Sir John Pakington, in which he argued that the time had come for a massive reconstruction of the British navy. It is an important but unpublished document and thus deserves lengthy citation:

> I beg to call their Lordships' attention to a subject of the highest importance, not only as regards the maintenance of the supremacy of the British Navy, but even the safety of the country; although, as I have frequently stated, it is not to the interest of Great Britain, possessing as she does, so large a navy, to adopt any important change in the construction of ships of war, which might have the effect of rendering necessary the introduction of a new class of very costly vessels, until such a course is forced upon her by the adoption by foreign powers of formidable ships of a novel character, requiring similar ships to cope with them, yet it then becomes a matter not only of expediency, but of absolute necessity. This time has arrived. France has now commenced to build frigates of great speed, with their sides protected by thick metal plates, and this renders it imperative for this country to do the same without a moment's delay. Having anticipated for some considerable time past that such a step as this would become necessary, the attention of my department has been directed to the preparation of designs, and the mode of construction of vessels of this novel character, and the design forwarded herewith, which was prepared several months ago, will show the general character and size of the vessels, which it is considered should be built as soon as possible to meet this emergency.[20]

Walker recommended, in addition to the conversion programme, that two ironclads be constructed in the government dockyards and four by private contractors,[21] further evidence of the British system of government–private cooperation in armaments brought on by a perceived crisis.

Sir John Pakington was 'mortified and vexed' by the report.[22] Signs of a *rapprochement* between France and Russia had appeared, which prompted the Admiralty to issue a warning, in March 1858, that British naval strength should be greater than the combined forces of France and Russia.[23] The Queen, upon her return from a visit to Cherbourg, the naval dockyard which the French were extending, added her voice to the alarm.[24] At the end of July the Admiralty, in a decision which underlined their determination to maintain naval superiority against any combination of enemies, resolved to build two ironclads which would be technologically superior to the French.[25] Efforts were made to learn more about the French ship in order to ascertain what specifications met that objective.[26] Tests began on various thicknesses of iron plate.[27] In the autumn, Derby appointed a Parliamentary Committee to study the naval balance between Britain and France and Walker's analysis of the situation was accepted despite its serious omissions.[28]

The Committee's recommendations to accelerate conversions of frigates to steam power were implemented by the government immediately. The naval estimates were set in the autumn of 1858 and in November Pakington took 'very strong representations' to the Cabinet which 'entirely acceded to the necessity of taking action as soon as possible'.[29] Thus the British Admiralty moved swiftly and decisively towards iron ships in response to the French, as they had done with steam in the 1820s and 1830s and screw propulsion in the 1840s.

In the winter of 1858–9, the Admiralty called upon private industry to submit designs for the new warship based on specifications derived from their tests the previous summer and information they were able to gather about the French ship. Tender invitations went out on 29 April 1859, the contract was awarded to the Thames Iron Shipbuilding Company and work began immediately. Under the impatient eye of the Admiralty, the company raced to catch up to the French, whose head-start was nearly a year. It was a close finish. The first British iron battleship, the *Warrior*, was launched on 29 December 1860, only a few weeks after the French ship, *La Gloire*, had finished its sea trials. Others ships of the *Warrior* class quickly followed: the *Black Prince* was launched on the Clyde in February 1861, with more over the next few years. French pursuit of a technological advantage over the Royal Navy was again frustrated by British industrial might and determination to retain the dominant power at sea. Paixhans' idea contained the same flaw as did that of the Tirpitz plan 40 years later. France and then Germany continued to seek naval equality with Britain through technological innovation.[30]

Along with its more speedy production from design to finished product, the *Warrior* was technologically ahead of *La Gloire*. Both ships carried 4½ inches of iron, but in all other respects the *Warrior* was superior. Firstly, it had an all-iron hull, whereas *La Gloire*'s armour was attached to a wooden hull.[31] The Admiralty's insistence that the *Warrior*

Note: This restored historic warship can be visited at the naval dockyard in Portsmouth.

Figure 2.1 Warrior, *National Maritime Museum, Greenwich*

be faster than *La Gloire* necessitated the iron hull. It had been learned that the French ship would do 12 knots. To exceed that, the *Warrior*'s hull had to be long and narrow: narrow to cut the water efficiently, long to accommodate the large engine to deliver the necessary power. A hull of this shape and strength could only be built with iron. Iron was also necessary to make the series of water-tight compartments at the unarmoured ends of *Warrior*, an advantageous trade-off between speed and weight of protective armour impossible in a wooden ship. Thus, while *La Gloire* was a structurally weak and tubby 252-foot ship, the *Warrior* was a strong, slim and graceful 367-foot machine which slid through the water two knots faster than its rival.[32] Remaking British naval power to maintain sea supremacy was expensive, and would become increasingly expensive as the new technologies developed. The *Warrior* cost an unprecedented £355 000 (excluding guns), but was considered a small price to pay to maintain naval supremacy.[33] It also made all existing battleships obsolete, and launched a naval revolution.

V

Warrior's 4½ inches of defensive protection confounded gun makers. The results of tests of plates and guns in the summer of 1858 were announced in August by Captain R.S. Hewlett of the gunnery training ship, *Excellent*. Four inches of wrought iron had withstood the blows from the navy's most powerful gun, the 68-pounder, fired from a distance of only 100 yards.[34] The artillery experts at the War Office scrambled to respond and in September they convinced General Jonathan Peel, the Secretary of State for War, to set up a Special Committee on Rifled Ordnance to investigate the state of British artillery and to suggest ways to improve it.[35]

The secretary of this committee was Andrew Noble, a young artillery captain who had studied ballistics. Several Armstrong breech-loaders were among those tested and the two men naturally met. Armstrong was impressed by Noble's knowledge of ballistics and invited him to join the Elswick firm. Noble was concerned about giving up a successful career in the Royal Artillery Corps, but in 1860 he resigned and moved to Newcastle. He remained with Elswick for the rest of his long life and became a major contributor to the success of the company: he developed a slow-burning powder in the late 1870s; played an active role in the tremendous expansion of the firm in the 1880s and 1890s; succeeded Armstrong as chairman upon the latter's death in 1900; firmly controlled the company during the great naval race with Germany in the 1900s; and retained his position until his death in 1915.

Peel's committee examined the guns of seven inventors. Five were discarded immediately, leaving only Armstrong and Sir Joseph Whitworth.[36] Whitworth had been developing projectiles and did not submit an entirely new gun. His 68-pound muzzle-loader was a standard naval gun rifled in the Whitworth manner. It was a worthy but not exceptional weapon. Armstrong, however, had made substantial progress on his breech-loaders. He had 'built-up' his original 3-pounder into 12-, 25- and 40-pounder field guns, a few of which he had already sold to the army. Their performance caused great excitement. The Commander-in-Chief of the army, the Duke of Cambridge, declared that the Armstrong gun 'could do everything but speak'.[37] On 16 November the Special Committee recommended the 'immediate introduction of guns fixed on Mr. Armstrong's principle'.[38]

The Admiralty also were much impressed by Armstrong's gun, especially the 40- and 70-pounders and the less-developed but promising 110-pounder. In March 1858, they had written to the War Office and requested they be informed of upcoming gun trials at Shoeburyness so that naval officers could attend.[39] The Derby Committee recommended that no new ships be built, since the Armstrong gun might well 'supersede the use of ordinary ships' guns, and possibly affect even the size and structure of ships of war'.[40]

In the winter of 1858–9, the Board of Admiralty accepted the advice of Hewlett, their expert on these matters, and requested 'in the strongest manner' that the Secretary of State for War supply them with a 'large number' of 40-pounder and 70-pounder Armstrong guns. The *Warrior* was the first ship to carry Armstrong guns. It had ten 110-pounders and four 70-pounders, along with its twenty-six 68-pounders (still the most destructive at short range). This was an enormously important order for Armstrong. The 110-pounders sold for £750 each, the 70-pounders cost £500 each. Thus each ship of the *Warrior* class carried £9500 of Armstrong guns in addition to the carriage, projectiles and fuses for each gun, the costs of which exceeded the cost of the gun itself. Thus each ship carried £20 000 or more worth of Armstrong weapons. Armstrong's Elswick factory manufactured half of these, Woolwich the other half. These orders not only gave his business a good start but established his reputation throughout the world as the man whose guns were aboard the historic ship *Warrior*, the most powerful ship in the greatest navy in the world.

VI

In 1858, Woolwich Arsenal did not make built-up, breech-loaded guns.

As Armstrong later recalled, his guns 'could not be obtained from any source except from the Elswick Works, in which I was a partner'.[41] Armstrong could have sold his new weapons on the international market and it was this that the British government was anxious to avoid. They therefore approached Armstrong 'in order to ascertain the terms on which he would be willing to enter into an engagement with the Government'.[42] The Secretary of state for War, General Peel, announced that, 'having ascertained the superiority of the gun, the Government could have no hesitation in at once doing everything in their power to make themselves masters of it [and he] had not the slightest hesitation in saying that there was hardly any sum which the Government would have felt themselves justified in refusing for those patents'.[43]

Armstrong had been discussing the matter with the War Office since November 1858, when the War Office had ordered two 12-pounder field batteries with carriages. 'Foreseeing the probability of their coming into use', Armstrong and his partners in the Elswick Engine Works had spent £12000 to erect a moderate-sized gun shop beside his crane factory. Further capital investments were necessary to accommodate new orders and Armstrong explained that, before he proceeded, 'some understanding with the Government should be come to on the subject'. Armstrong proposed the following: he would *give* the government his patents and furthermore agree not to sell his guns to any foreign nation, if the government would guarantee that the capital investment made by himself and his partners in the Engine Works would not be lost if the government decided to build all the guns at Woolwich or contract them out to others. The deal was completed on 15 and 16 January 1859, when a ten-year contract was signed. By then the naval orders were coming in and the government guaranteed £50000 in capital. This was soon raised to £85000 as the Elswick Engine Works rapidly expanded into armaments production.[44]

The contract satisfied both parties. The government were confident that the new gun would remain a British secret. The Company was bound not to disclose any information about the invention beyond what 'shall be necessary to enable the person or persons so employed efficiently and properly to conduct and carry on and complete the manufacture of the said guns'. New partners could only join the firm if they entered into the covenant. The government reserved the right of first refusal on loans the company might seek, 'it being the special design and object of this stipulation to avoid as much as possible the introduction of any strangers into the transactions connected with the present contract in order that the risk of the disclosure of the said inventions may not be unnecessarily incurred'.[45] Finally, Armstrong transferred his patent rights to the Secretary of

State for War and the government amended the patent laws to ensure absolute secrecy.[46]

The agreements signed on 15 and 16 January 1859 created a symbiotic relationship between Woolwich and Elswick. The guns were produced in both factories and thus the two establishments were in competition with each other for a single customer, the British government. Woolwich was the government's factory but it was obliged to buy guns from Elswick 'sufficient to keep the works erected and to be erected as aforesaid in full and adequate employment'. Further research costs were borne by the government and not included in the price of the guns. Government approval was required for all experiments and trials. Prices were to be 'fair and reasonable'. Disputes unresolved by negotiation would go to the Attorney General for arbitration.

Armstrong was appointed Engineer of Rifled Ordnance at Woolwich, which allowed him to supervise the manufacture of his guns there and, more importantly, continue his research and development at government expense. He received £2000 per year salary, backdated to 1856 in order to compensate him for the costs that he had incurred during those years. Finally, Armstrong was given a knighthood from a grateful nation.[47]

Thus the contract between Armstrong, the Elswick Ordnance Works and the British government delicately balanced the interests of the state and the business interests of Armstrong and his partners. Each side in the agreement had clear obligations and privileges, but Armstrong benefited most. The business was guaranteed and he had the entire ordnance department at Woolwich at his disposal to pursue his research and development.

The Elswick factory was separated from the civil engineering business and named the Elswick Ordnance Company. Its board of directors were Armstrong's partners in the Elswick Engine Works: George Cruddas, Richard Lambert and George Rendel, son of James M. Rendel, who had died in 1857. Cruddas took responsibility for financial control and George Rendel, who had studied engineering and lived with Armstrong after his father died, managed the factory. Armstrong was not a legal partner, but the firm depended on his invention and his substantial loan. He reserved the right to join the firm at any time. Meanwhile he continued his research at Woolwich.

On 4 March, Peel's announcement that an agreement had been signed was enthusiastically received in Parliament. He proclaimed that Armstrong guns would sooner or later 'supersede the whole existing armament of the country' and that Mr Williams, a leading opponent of armaments expenditures, would have to 'make up his mind to a large expenditure on that account'. Williams agreed with a 'Hear, hear.'[48] Sidney Herbert, who would soon be the Secretary of State for War in a

Liberal Cabinet, gave full credit to the Tory government for getting the Armstrong gun, and proclaimed that the gun was so 'thoroughly proved', that the government was 'justified in waiving the usual caution not to spend too much money in the manufacture of a newly invented weapon lest further improvements should be found to supersede it in its present condition'.[49] General Codrington, hero of Alma and Inkerman, and Liberal M.P. for Greenwich, 'rejoiced to hear that energetic measures would be adopted for introducing the Armstrong gun into service' and heartily agreed that England should have, 'at whatever cost, the most destructive warlike weapons that could be invented'.[50] Thus, by the spring of 1859, Armstrong's career had taken a dramatic shift from that of a civil engineer and businessman to a virtual national hero.[51] The fortunes of Elswick were tied to the British army and navy, and all seemed set for the company to become a major industrial player in a new kind of military–industrial complex.

The rebuilding of the British navy meant business for the British iron industry in general and firms which supplied the British navy in particular. Marine steam engines, iron plates, iron hulls and all sizes of guns were in heavy demand. These industrial products were expensive. In 1859, the conversion of one battleship to steam cost about £25 000; a new steam battleship was more efficient but cost £105 000. In 1851, the navy spent less than £100 000 on steam engines; by the end of the decade the average amount was £570 000 per year. Between 1852 and 1858, the number of steam vessels rose from 177 to 464 and the number of guns from 3045 to 8246.[52] In the few years before the Crimean War the army budgets (which included all naval ordnance) and naval budgets had average totals of about £9 million and £5 million, respectively. They more than tripled during the war but did not return to their pre-war levels afterwards. The army budget rose from £14 million in 1860 to £17 million in 1863 and settled around the £15 million mark throughout the remainder of the decade. Naval spending increased from £8 million in 1859 to £10 million in 1860 and to £13 million in 1861, before receding to under £10 million in 1864.

The Elswick Ordnance Company was formed when these budgets were undergoing their sharpest climbs. Unlike the ship builders and engine makers, Armstrong's firm was new and required high capital expenditures, and any extra demand for guns would get the company off to a good start. The January 1859 contract to supply the government with Armstrong breech-loaders was thus signed at a propitious moment. Five months later, French soldiers were in Italy, Palmerston was again the Prime Minister and the Secretary of State for War was Sidney Herbert.

Demand for Armstrong guns surged in July 1859 as the invasion panic reached its height. Production could not keep up with demand, thus throwing fuel on the invasionists' fire. Only 100 Armstrong guns would be ready by Christmas and just 300 by the spring of 1860, but Herbert assured Parliament that the government was pushing production as hard as it could. Only a short time ago, he said, he had himself asked Sir W. Armstrong whether any sum of money which would enable him to procure additional labour or more rapid machinery would assist him in the quicker production of those guns.[53] But Armstrong had as many workers and as much machinery as were available and production was running at full capacity in both the Woolwich and Elswick factories.[54] By November, Herbert had stopped all production of artillery at Woolwich and had placed the entire Royal Gun Factory under Armstrong who supervised the manufacture of his guns.[55]

Elswick received half the orders. In the fiscal year 1858–9, the company had produced fewer than £8000 in guns, projectiles, carriages and their appurtenances. In 1859–60, £31 500 worth of weapons was ordered by the War Office and Armstrong expanded the factory as quickly as possible. In each of the next two fiscal years the company's revenues reached £400 000.[56] Between 1858 and 1863, Elswick and Woolwich combined produced £2 560 000 worth of Armstrong breech-loading guns in six sizes: 6-, 12-, 25-, 40-, 70- and 110-pounders. Elswick's share was £1 525 000.[57]

Costs and prices fell over the first year or two as Armstrong bought new machinery. In 1858, Armstrong's estimate for a 110-pounder was £1200, but within two years it was down to £700 and in 1862 the price was £600. The price of the 25-pounder was £220 and the 12-pounder cost the government £140.[58] Ammunition also became cheaper. The price of a 110-pound shot fell from £1 5s to under £1. The 40-pound shot sold for about 10s, the 12-pound for 5s and the 6-pound for 2s. The government bought many more shells than shot and they ranged in price from £1 for the 110-pound down to 2s for the 6-pound. Fuses cost between 1s and 6s.[59]

Within five years from the time Armstrong produced his first 3-pounder gun in 1855 he was manufacturing a full range of guns on the same built-up principle. His factory at Elswick was producing hundreds of guns per year and he was supervisor of the government factory at Woolwich, where he continued his research and development into making larger and more powerful guns. His success was the result of his own engineering genius and the historical circumstances of the 1850s in which war and the threat of war ensured that the demand for guns was strong. Thus by the spring of 1859 Armstrong's career had taken a dramatic shift, from that of a civil engineer and businessman to a virtual national hero.[60] The fortunes of Elswick were tied to the British army and navy, and all seemed set for the company to become a major industrial player in a new kind of military–industrial complex.

Notes

1. Martin van Creveld, *Technology and War: From 2000 B.C. to the Present* (London: The Free Press, 1989), pp. 132–4.
2. Theodore Ropp, *War in the Modern World*, revised edn (New York: Collier Books, 1962), p. 51.
3. Tom Hartman, *The Guinness Book of Ships and Shipping: Facts & Feats* (London: Guinness Superlatives Limited, 1983), p. 74.
4. Peter Kemp (ed.), *The Oxford Companion to Ships & The Sea* (London: Oxford University Press, 1976), under 'Naval Guns'; A.R. Hall, *Ballistics in the Seventeenth Century: A Study in the Relations of Science and War with Reference Principally to England* (New York: Harper & Row, 1969), pp. 168–71.
5. Ropp, *War in the Modern World*, p. 72.
6. There are many naval histories of the period between 1500 and 1800. The above outline is based on two books by Antony Preston: *History of the Royal Navy* (London: Hamlyn, 1983) and *Battleships* (New York: Cresset Books, 1981); see also Frederick Robertson, *The Evolution of Naval Armament* (London: Constable & Company Ltd., 1921); Christopher Lloyd, *The Nation and the Navy: A History of Naval Life and Policy* (London: The Cresset Press, 1954). For a broader view of the growth of British naval power, see Paul Kennedy, *The Rise and Fall of British Naval Mastery* (London: Allen Lane, 1976).
7. Michael Lewis, *The History of the British Navy* (London: George Allen & Unwin Ltd, 1959), p. 199.
8. L.T.C. Rolt, *Victorian Engineering* (Harmondsworth, Middlesex: Pelican Books, 1974), pp. 85–8.
9. James P. Baxter, *The Introduction of the Ironclad Warship* (Cambridge, Mass.: Harvard University Press, 1933), pp. 36–8; Antony Preston, *History of the Royal Navy*, pp. 104–5.
10. Theodore Ropp, *The Development of a Modern Navy: French Naval Policy 1871–1904*, Stephen S. Roberts (ed.) (Annapolis, Md.: Naval Institute Press, 1987), p. 219.
11. Baxter, *Introduction of the Ironclad*, p. 77.
12. *The Times* (16 Dec. 1854, p. 12) gave credit for the idea to the French, but the Admiralty took a more defensive stand. After the war, Sir Charles Wood dismissed French technological leadership and emphasised England's greater productive contribution: 'Report from the Select Committee on the Board of Admiralty', *Parl. Papers* V (1861), qs. 2763–6.
13. Baxter, *Introduction of the Ironclad*, pp. 117–18.
14. Ibid., p. 118.
15. 'Report on the Board of Admiralty', *Parl. Papers* V (1861), qs. 2761, 2886.
16. Preston, *History of the Royal Navy*, p. 103.
17. Biographical information about Walker is from *D.N.B.* and Burke's *Peerage*. The reasons for his appointment as Surveyor are provided in 'Report on the Board of Admiralty', *Parl. Papers* V (1861), qs. 2334, 3137–9 and *Hansard*, 28 Feb. 1861, col. 1136.

18. *Parl. Papers* V (1861), qs. 2508–12.
19. Andrew Lambert, *Battleships in Transition: The Creation of the Steam Battlefleet 1815–1860* (Annapolis, Md.: Naval Institute Press, 1984), p. 38.
20. 'Report on the Board of Admiralty', *Parl. Papers* V (1861), q. 1703.
21. Ibid.; Baxter, *Introduction of the Ironclad*, p. 123.
22. 'Report on the Board of Admiralty', *Parl. Papers* V (1861), q. 1706.
23. Baxter, *Introduction of the Ironclad*, pp. 119–20; Kenneth Bourne, *The Foreign Policy of Victorian England 1830–1902* (Oxford: Clarendon Press, 1970), pp. 78–80, 96–8, 100, 114.
24. Andrew Lambert, *Battleships in Transition: The Creation of the Steam Battlefleet 1815–1860* (Annapolis, Md.: Naval Institute Press, 1984), p. 71; Maurice Pearton, *The Knowledgeable State: Diplomacy, War and Technology Since 1830* (London: Burnett Books, 1982), p. 57.
25. 'Report on the Board of Admiralty', *Parl. Papers* V (1861) q. 1706; Baxter, *Introduction of the Ironclad*, pp. 123–4.
26. 'Report on the Board of Admiralty', *Parl. Papers* V (1861), q. 1287.
27. Baxter, *Introduction of the Ironclad*, pp. 126–8.
28. French ships used inferior timber whose ability to withstand the extraordinary stresses from the weight of their steam engines was doubtful. Moreover, the ratio of French paddle to screw frigates was higher than that of the British fleet. See Lambert, *Battleships in Transition*, pp. 75, 77–8.
29. 'Report on the Board of Admiralty', *Parl. Papers* V (1861), qs. 1287–8.
30. French and German naval authorities saw the torpedo as the piece of technology which would eliminate British battleship superiority. The young Tirpitz wrote of the torpedo as 'a weapon ... as important as Paixhans's shell' (Ropp, *Development of a Modern Navy*, p. 134).
31. The French iron-hulled ship, *La Couronne*, was begun a year before the *Warrior*, but launched a month after the British ship.
32. *Warrior* and *Black Prince* were known as the 'black snakes of the Channel'.
33. 'Total Cost of H.M. Ship "Warrior" Before Being Ready for Sea', *Parl. Papers* XXXIV (1862), p. 895. Technical specifications are in Oscar Parkes, *British Battleships, 'Warrior' 1860 to 'Vanguard' 1950: A History of Design, Construction and Armament* (London: Seeley Service, 1966) and *Conway's All the World's Fighting Ships 1860–1905* (London: Conway Maritime Press, 1979).
34. Baxter, *Introduction of the Ironclad*, pp. 123–6.
35. *Confidential Report*, W.O. 33/9; Select Committee on Ordnance, *Report* (1863), p. iv; Sir J. Emerson Tennent, *The Story of the Guns* (London, 1864), p. 125–6.
36. 'Report from the Select Committee on Ordnance', *Parl. Papers* XI (1863), p. iv.
37. J.D. Scott, *Vickers: A History* (Weidenfeld and Nicolson, 1962), p. 28.
38. W.O., *Confidential Report*.
39. Navy Board Minute, 30 March 1858, Adm. 3/266.
40. 'Report of a Committee of the Treasury to Inquire into the Navy Estimates, from 1852 to 1858, and into the Comparative States of the Navies of England and France' (Derby Committee), *Parl. Papers* XIV, Part 1 (1859).
41. Armstrong, Evidence to Royal Commission on Warlike Stores, *Parl. Papers* XV (1887), q. 9265. In the same testimony, Armstrong more accurately referred to himself as the 'principal acting partner' in the Elswick firm.

42. W.O. *Confidential Report*, W.O. 33/9, p. 813.
43. *Hansard*, 4 March 1859, col. 1319.
44. The negotiations are in Appendix 11 of the 'Report of the Ordnance Select Committee', *Parl. Papers* VI (1862) and in the testimony of that report, q. 2339–51. A copy of the contract signed on 16 January is in the 'Report of the Committee on Military Organization', Appendix 10, *Parl. Papers* VII (1860), pp. 723–7. See also *Hansard*, 11 March 1859, col. 38.
45. All quotes are from the contract cited in the previous footnote. See also Stuart Rendel, *Personal Papers of Lord Rendel*, F.E. Hamer (ed.), p. 271; Scott, *Vickers*, p. 28; Dougan, *Great Gun-Maker*, p. 64; McKenzie, *W.G. Armstrong*, p. 68; remarks by Secretary for War, Peel, *Hansard*, 4 March 1859, col. 1319; *D.N.B.*, vol. 22 (Supplement), p. 66; *Encyclopedia Britannica* (1972), vol. 2, p. 443.
46. The Bill was introduced on 21 March and passed on 11 April. See *Hansard*, 21 March 1859, col. 482; *The Times*, 12 April 1859, p. 7; and Armstrong's testimony to the Royal Commission on Warlike Stores, *Parl. Papers* XV (1887), q. 9265.
47. The two agreements of 15 and 16 January are in the 'Report of the Committee on Military Organization', *Parl. Papers* VII (1860), pp. 723–7. Armstrong's job description as Engineer for Rifled Ordnance is in W.O. 33/9.
48. *Hansard*, 4 March 1859, col. 1320.
49. Ibid., cols 1322, 1325.
50. Ibid., col. 1332. See the *D.N.B.* for Codrington's background.
51. *The Times*, 26 Feb. 1859, p. 5.
52. 'Report of a Committee of the Treasury to Inquire into the Navy Estimates, from 1852 to 1858, and into the Comparative States of the Navies of England and France' (Derby Committee), *Parl. Papers* XIV, Part 1 (1859), pp. 703, 712.
53. *Hansard*, 15 July 1859, cols 1323–4.
54. Ibid.
55. Stanmore, *Sidney Herbert*, vol. II, p. 210.
56. 'Statement Showing the Payments to the Elswick Ordnance Company', *Parl. Papers* XI (1863), p. 560.
57. 'Statement Showing Total Amount Expended in the Manufacture of Guns and Projectiles on Sir William Armstrong's Principle', *Parl. Papers* XI (1863), p. 558. It is worth emphasising here that 'guns on Sir William Armstrong's Principle' refers only to the breech-loaders made on the built-up principle. As we shall see in the next chapter, Armstrong also began to develop large muzzle-loaders.
58. 'Statement Showing the Amount and Rate Paid for certain Guns to the Elswick Ordnance Company Compared with Original Rates', *Parl. Papers* XI (1863), p. 514.
59. 'Statement Showing the Cost of certain Guns and Projectiles Obtained from the Elswick Ordnance Company', *Parl. Papers* XI (1863), pp. 558–60.
60. *The Times*, 26 Feb. 1859, p. 5.

CHAPTER THREE

Defending England: monster guns or iron ships?

I

Armstrong's built-up gun contributed to a fundamental change in British defensive strategy in the 1860s and in British warship development during the rest of the century. As had happened in the fourteenth and fifteenth centuries, artillery innovations in the nineteenth century upset traditional military assumptions and, when new guns were adopted by navies, new kinds of warships were developed to carry them. Sir William Armstrong became an important figure in the nineteenth-century phase of this historical process. The military and naval decisions made by the British government in response to steam-powered, ironclad warships were closely related to and, indeed, dependent upon Sir William Armstrong and his firm belief that he would build larger and more powerful guns.

Steamships brought into question the ability of the British navy to fend off invasion and consequently provided an opportunity for the British army to expand its place in British defence strategy. Until 1859, the army had been considered the protector of British colonial possessions and, along with the navy, defender of British interests in the world at large. The bombardment and invasion of Beirut and Acre in 1840 was repeated, albeit with more difficulty, on a larger scale in the Crimea in 1855. The army added to its reputation by putting down the Indian Mutiny of 1857 and returning the British kind of law and order to China in 1858.

The war in Europe and the tactical uncertainties of British naval strategy created by steam and iron in 1859 opened the way for the British army to claim a new and prestigious role as the ultimate defender of English shores. Shortly after taking office as the Secretary of State for War in Palmerston's Liberal government, Sidney Herbert proposed that a fundamental change be made in British defence strategy, explaining to the Cabinet, 'We must look at our Army now altogether from a different point of view. Our insular security as such is lost. No mere preponderance of our fleet in the Channel can insure [*sic*] perfect safety.'[1] Herbert believed that the popular expression that steam had bridged the Channel was misleading. A bridge was fixed in one position and invaders using it could be kept out by a few well-placed batteries, but steamships could cross virtually anywhere and only a large army and extensive coastal fortifications under army command could defend against troops landed by these new ships. Herbert was convinced that the Royal Navy could no longer guarantee British national security.[2]

Palmerston, who had made the bridge metaphor popular in the 1840s, set the political machinery in motion to channel extra millions into land defences. He created a Cabinet Defence Committee consisting of himself, Herbert, the Duke of Cambridge (Commander-in-Chief of the army) and Lord Somerset (First Lord of the Admiralty), who supported building forts to protect the naval dockyards and arsenals. In August 1859, he appointed a Royal Commission on the Defence of the United Kingdom to study the technical requirements. The commissioners, however, were not impartial investigators of British defence strategy. Palmerston and Herbert did not seek an objective analysis of a defensive system based on forts and one based on the navy. Despite the importance of the question of British defence and the large amount of money involved in its recommendations (£12 million), it was one of the shortest Royal Commissions in British history.[3]

The conclusions and recommendations of the commissioners were predetermined in two ways. Firstly, the terms of reference excluded an investigation into the merits of the fortifications programme itself but confined the inquiry to making technical recommendations for the implementation of the policy. Furthermore, Herbert instructed the commissioners to study forts to protect dockyards and arsenals.[4] There was to be no discussion about the defence of London or any other part of Britain, yet the Commission report rested on the assumption that 'the nation cannot be considered as secured against invasion if depending for its defence on the fleet alone'.[5] The Cabinet Defence Committee also thought that protecting the arsenals and dockyards would 'check the alarms caused by fear of invasion, which from time to time take possession of the public mind'.[6] Financial and technological arguments against the forts were brushed aside.[7]

The recommendation to build forts was assured, secondly, by the choice of the members of the Commission. Each of the five army officers and the two naval officers and the one civilian had a personal commitment to fortifications. One of the naval members, Sir Ashley Cooper Key, had delivered a paper to the Admiralty in January 1859, in which he argued the case for forts to protect the naval arsenals from sudden attack.[8] Sir George Elliot, the other naval Commissioner, had been impressed by the fortifications at the siege of Cadiz in 1811.[9] The civilian member of the Commission, James Fergusson, had argued on behalf of fortifications since the invasion scares of 1848 and 1852–3 with periodical articles such as 'The Perils of Portsmouth' and 'French Fleets and English Forts'.[10] Ignored for years, Fergusson suddenly found himself on the 1859 Commission, where his ideas were taken seriously by powerful people. The chairman of the Commission was Sir Harry Jones, whose long career had been devoted to building forts. His first assignment after engineering school was at the extensive fortifications at Dover and he too was at Cadiz and other sieges during Wellington's Spanish campaign. After Waterloo, he became Inspector of Fortresses in Europe and he witnessed the failed siege on the Redan in the Crimea in 1855. He wrote a number of works on sieges and fortifications and, after his work on the Royal Commission was completed, promotions and awards flowed his way.[11]

Sir William Jervois, the secretary to the Commission and the man who was most responsible for writing its report, was also a strong advocate of fortification. He had been Assistant Inspector-General of Fortifications at the War Office in 1856, secretary to the Defence Committee under the Duke of Cambridge in 1857, and later an advisor to successive secretaries for war on matters of defence. As governor of South Australia in the 1870s, he carried out a number of coastal defence projects and in the 1880s he undertook a similar programme in New Zealand. 'He was much aided in these endeavours by the war scare in 1885.'[12]

The commissioners wanted to build forts that were safe from guns fired from (French) ironclads. It was up to Sir William Armstrong to supply the forts with guns which would sink invading ironclads. Armstrong, however, did not yet have such a gun. His 110-pounder had a maximum range of 9175 yards, but its effective range was much less. At 8000 yards, a target area of 5000 square yards was hit by only about one of every three shells fired. At 4000 yards, a small gunboat was quite safe. His time fuses were not reliable for long distances, and he did not yet have a gun carriage for this type of battery.[13] But, most importantly, the 110-pounder could not penetrate iron plate at even close range. Despite these realities, the government were determined to build the forts. The passage of the Bill through Parliament was carefully planned. It was not mentioned in the Queen's speech or in the budget. Herbert submitted the £12 million Fortifications Bill on 23 July 1860 and pushed it through as quickly as possible.[14] Construction began immediately at all the dockyards and at Spithead.

Those opposed to the forts continued their campaign. In 1861, Captain Cowper Coles, a strong advocate of naval power and turret ships, convinced Admiral Richard Dundas, the senior naval lord, that the Spithead forts would be ineffective against an iron turret ship. Coles gathered substantial support and Herbert reconvened the Royal Commission on Fortifications. The commissioners invited written presentations from Sir John Burgoyne, the Inspector-General of Fortifications, and Armstrong. Both men supported the fortifications and the programme continued.[15]

II

A naval engagement took place in the American Civil War which cast serious doubt upon the forts. On 8 March 1862, an ironclad steamship of the Confederate navy, the *Merrimac*, came under close-range fire from the guns of a Union shore battery at Hampton Roads, Virginia, at the mouth of the James River. The *Merrimac* shrugged off the projectiles fired from the shore batteries while she turned her shell guns on the Federal wooden sailing frigate *Congress*, setting her ablaze. The following day, the *Merrimac* returned to finish off several other ships and break the Federal naval blockade but was met by the Federal ironclad turret ship, *Monitor*, which had arrived on the scene overnight. The *Merrimac* was a broadside ship (its 7-inch and 9-inch guns were ranged along the sides in the traditional arrangement); the *Monitor*'s two 11-inch shell guns were in a rotating turret. The *Merrimac* was far larger than the *Monitor*, but the latter ship drew less water, was more manoeuvrable, presented a much smaller target and had larger guns. On the morning of 9 March the first clash between two ironclads took place. The two ships hammered each other at point-blank range for four hours, with no effect on either. The indecisive outcome of this battle, which contrasted so sharply with the damage inflicted by the *Merrimac* against wooden ships the previous day, proved to many that ironclads were impenetrable by even the most powerful guns at the closest range.[16]

This incident undermined support for Palmerston's forts. On 31 March, Members of Parliament confessed one after another that the performance of the *Merrimac* had changed their views about the forts. Sir Frederick Smith had previously voted against a motion to abandon the Spithead forts 'because at the time he thought such measures of defence were necessary there; but recent events had induced him to alter his opinion'.[17] He calculated that it was not too late for them to cut their losses on the forts,[18] and he called for a return to the naval defence advocated by his 'lamented friend' Sir Richard Dundas. Ships, not forts, would protect England from invasion.[19]

Others pointed out that Armstrong did not have a gun that could penetrate the 4½ inches of plate carried by *La Gloire*.[20] French or American ironclads had to be met up close, with British ironclads carrying the old 68-pounder muzzle-loader, the most powerful gun available at short range (200 yards). Sir John Hay, who had been secretary to the Royal Commission on Defence in 1859, confessed that he too had changed his mind and now advocated a return to naval defence.[21]

Sir George Lewis, who had taken over as Secretary of State for War after Herbert's death, defended the forts as best he could. Ironclads, he said, were more expensive than forts, and he claimed a supplementary estimate of £10 million or £15 million would be needed to build an ironclad fleet. But Hampton Roads had created a problem for the government: 'I do not say [declared Lewis] that it may not be necessary, in consequence of the experience of the recent engagement, to make a vast change in our naval defences.'[22]

The secretary of the Admiralty, Sir Clarence Paget, defended a policy which reduced the role of the navy. He admitted that 'naval men naturally prefer ships to forts' – he did not add that military men naturally prefer forts to ships – and proclaimed that 'there can be no doubt that forts must ever be stronger than ships'. He insisted,

> nothing which has taken place across the Atlantic in any way alters the position of the question. Let it be remembered that as regards ships, there must be a limit ... to the size of the gun and the thickness of the plate which they can carry. But in a fort there is no limit. If we can construct a gun to carry a shot of 1,000lb weight we can mount it on a fort.[23]

Tests had repeatedly demonstrated that ships with 4½ inches of armour, like *La Gloire* and *Warrior*, were impregnable to all existing shot and shell. Neither the Armstrong 110-pounder nor the 68-pounder muzzle-loaded smooth bore could penetrate iron of even two inches.[24] Ironmasters seemed to have secured a permanent advantage over gun manufacturers and they guarded their reputation closely. When the Duke of Somerset claimed that a shot from an Armstrong gun had penetrated eight inches of iron in 1861, the Earl of Hardwicke protested that this was impossible; the Armstrong gun had never penetrated 4½ inches of plate and to claim otherwise was 'calculated to be highly injurious to the iron manufacturers' of England.[25]

On 4 April 1862, a motion to suspend construction of the Spithead forts and redirect the money to the construction of new ironclads passed.[26] The entire forts programme was vulnerable to a similar vote unless Armstrong made some tangible progress towards a gun that would penetrate armour.

III

The two Spithead forts overlooked 2400 yards of water and hence the guns in each had to penetrate 4½ inches of iron at a range of 1200 yards in order to guarantee that an ironclad like *La Gloire* would come within effective range of the guns in one or the other fort. By 1860, Armstrong realized that his breech-loaders could not be made strong enough to send a projectile through iron at this distance. The explosion required to impart the necessary velocity to the shot or shell also destroyed the breech mechanism. He thus began work on a second generation of built-up guns which were loaded through the muzzle, and had soon produced rifled muzzle-loaders that weighed 6, 12 and finally 22 tons. His largest breech-loader, the 110-pounder, weighed only 4 tons and the navy's 68-pounder 4¾ tons. Armstrong's 12-ton gun fired a 150-pound solid shot or a 300-pound elongated shell; his 22-ton gun threw a 600-pound shell. They were called 300-pounders and 600-pounders (or 'Little Will' and 'Big Will' after their maker). These enormous guns (sometimes called 'monster' guns) used between 60 and 140 pounds of powder, while the 110-pounder breech-loader used between 10 and 14 pounds.

Armstrong adopted a method of rifling muzzle-loaders first developed by a French colonel in 1842.[27] Studs attached to a shell guided it down the barrel on one set of grooves and then it was shunted into another set for firing, thus giving these rifled muzzle-loaders yet another name: 'shunt guns'. Round shot had greater penetration than shell at short ranges (up to 200 yards) but the elongated shell had greater range and accuracy. It was a matter of muzzle velocity and wind resistance during flight. A shot could be given a higher muzzle velocity than the heavier shell but lost momentum much more quickly than the elongated, spiralling shell. Shot had the best chance of penetrating an ironclad at short range (and hence had to be fired from ships which could carry them into range); the heavier shells had greater range and accuracy but lacked the momentum at impact to penetrate iron. It was Armstrong's task to build a rifled muzzle-loader whose shell would pierce at least 4½ inches of iron from as great a distance as possible. Supporters of the forts believed that such a gun would be too heavy to be carried by a ship (Armstrong, it must be noted, did not agree with this assumption). Armstrong hoped to penetrate the *Warrior* target first with a smooth-bore 150-pounder from 200 yards, and then with a rifled 600-pounder from 1200 yards.

Armstrong completed his first 300-pounder in the spring of 1862 and delivered it to the government firing range at Shoeburyness on the north bank of the mouth of the Thames. The test, held on 9 April, was observed by an illustrious audience: the Duke of Cambridge on behalf of the army, the Duke of Somerset and Lord Clarence Paget from the Admiralty, Sir John Hay of the Special Committee on Iron Plates (made up of military, naval and civilian experts, including Sir William Fairbairn, the ironmaster who made the plates), Samuda the ship builder and, of course, Sir William Armstrong. Behind them stood many lesser

War Office and Admiralty witnesses. It was not a public event, however; *The Times* noted that it was conducted with 'unusual privacy',[28] and two months later the government defeated a motion which called for the results to be published.[29] However, as Sir George Lewis admitted, 'those who [took] special or professional interest in the subject had no difficulty in obtaining accurate information as to the result'.[30] The day after the test, *The Times* published an enthusiastic and dramatic account of the event. The gun, the editor wrote, 'was tried against that great champion of heavy weights which has hitherto come off victorious in all encounters – the redoubtable Warrior target'. After several shots, all doubts were resolved about the new champion:

> With an indescribable crash that mingled fearfully with the report of the gun, the shot struck upon a comparatively uninjured plate, shattering the iron mass before it into little crumbs of metal, splintering the teak into fibres literally as small as pins.

Another shot, with a larger charge behind it, drove completely through the target and

> even the fondest believers in the invulnerability of our present ironclads were obliged to confess that against such artillery, at such ranges, their plates and sides were almost as penetrable as wooden ships. ... Admiralty officials and armour shipbuilders could only admit to each other in a kind of confidential dismay, that artillery had at last proved too much for them.[31]

The government was not displeased that *The Times* had published a favourable report of a secret military test. Indeed, it effectively became the government's report. The Undersecretary of State for War (who would soon be the secretary after the death of Lewis), Earl de Grey and Ripon, exclaimed to his fellow peers that

> it would be superfluous of him to occupy their Lordships' time at any length on [an account of what he witnessed at Shoeburyness] after the very able and graphic description which had appeared in *The Times* newspaper, for the accuracy of which descriptions he was prepared to vouch.[32]

The government, in the midst of the debate over the Spithead forts, wasted no time using the newspaper report to defend its fortifications policy. De Grey proclaimed his faith that Armstrong would soon have a gun to put in the forts:

> Sir William Armstrong – a cautious man in speaking of anything which had not had the test of experience – had assured him that day, that although he would not speak with certainty on the point, he had no doubt in his own mind that a gun could be made which would be effective at the mean range between the two forts at Spithead.[33]

The Duke of Cambridge added his voice of authority. He too had 'communicated with that distinguished and able man Sir William Armstrong', who had expressed 'the strongest conviction' that he could make a rifled gun which would do at 2200 yards what could only be done at the moment at 200 yards. As the Duke pointed out that 'if we got only half of what Sir William Armstrong promises, or 1,100 yards, these Spithead forts would at once become invaluable'.[34]

Palmerston was a little more cautious. Twelve hundred yards was a very long way from 200 and he wrote to the queen that 'it yet remains to be seen whether or not penetration at 1200 yards was possible'. Nevertheless, he too had confidence in Britain's greatest gun maker, and confidently reported to Queen Victoria that 'Sir William Armstrong says he can and will make a gun that will send a shot which even at distance will pierce iron plating.'[35]

Armstrong would continue testing his monster guns and 'the value of forts placed in the proposed positions must depend very much upon the result of such experiments'. If, Palmerston explained, Sir William could make a gun to smash the sides of ironclads at 1200 yards, 'then there can be no doubt as to the usefulness of the proposed forts', but if he failed, 'some other arrangement' for defence of English arsenals and dockyards would be necessary.[36]

It was soon apparent, however, that the description of the gun's performance published by *The Times* was exaggerated. The shot had not passed completely through the iron skin behind the wooden backing of the target.[37] Moreover, the target used had been damaged by previous tests of a 68-pounder. The government was attacked for misrepresenting the results.[38] Sir John Hay admitted that they had been hasty in celebrating the results, but he tried to make the best of it:

> Whether by happy accident or as the result of the possession of better data than those on which the builders of other ships had gone, it so happened that those who constructed the *Warrior* had adopted the most satisfactory form of side for sustaining the effect of heavy projectiles.[39]

Palmerston waved aside the attacks and claimed that MPs should not concern themselves with the 'administrative details' of policy.[40] On 3 June 1862, a long and rowdy debate took place over guns and forts. In a packed House, and to everyone's surprise (and outrage), Palmerston made the issue one of confidence and the government's majority held. It was, Gladstone recorded, a complete rout.[41]

The subject emerged from time to time over the summer but there was no danger to the government, despite the fact that Armstrong's gun had not completely penetrated the *Warrior* target. Their faith in Armstrong did not weaken. As Sir George Lewis replied to one blistering attack on the forts:

> All that the Government can do is to consult a sufficient body of competent and disinterested scientific judges, and to adhere to the advice of that body when it is given ... I am not giving my own opinion. I wish it to be distinctly understood that I entirely decline to discuss questions of military science on any judgement of my own. I do not think that any advantage would accrue if we attempted in this House to debate technical matters of that kind except by a reference to scientific and experienced authorities.[42]

Palmerston agreed that in questions of military or naval tactics the government relied on others: 'all that the Government can do is to take the opinion of men whom they think fit and competent judges. That is what we did, and what we do'.[43]

Meanwhile Armstrong worked on his 600-pounder which would throw a shell through the *Warrior* target. The results of the 9 April test may have been exaggerated, but the shot had penetrated further than any previous one had done and he was confident that, if he could give a 600-pound shell the same velocity at impact the target would be destroyed.

IV

In September 1862, an engineering opponent burst on the scene, one with the engineering reputation, political support and newspaper backing to equal those of Armstrong. Sir Joseph Whitworth had been Armstrong's only competitor at the trials in 1858 arranged by General Peel, the Conservative Secretary of State for War, to determine the best kind of rifled artillery for the army. Only a few years earlier, Whitworth had been the government's principal consultant on rifled guns and called by the Commander-in-Chief of the British army 'the most celebrated mechanician' in the country.[44] He had been the government's closest advisor on rifled small arms during the Crimean War. Whitworth was famous for his precision machine tools – he was the first engineer to measure one millionth of an inch – and, two months after they had declared war on Russia, the British government invited him to improve the machines used to manufacture Enfield rifles. Like Armstrong, Whitworth was not a gun maker, but he too did not hesitate to take up a technological challenge. He convinced the government to finance a thorough investigation of rifling and projectiles. In 1856, the War Office began supplying Whitworth with large ingots to rifle into artillery pieces and he received £17000 in Government grants to carry out his research.

All ended in November 1858, when the Armstrong breech-loaders captured the imagination of the military authorities. Armstrong was celebrated as England's great gun maker and showered with government posts and honours and his new armaments business was given a monopoly to supply Britain with all breech-loaders not made at Woolwich. Whitworth's gun was rejected, all financial support for his research was withdrawn and Whitworth became another victim of the Armstrong built-up breech-loader.

Whitworth did not give up. Determined to prove that his system of rifling and his shells would penetrate armour, he returned to Manchester, continued his research, and in 1860 obtained partial success against the *Trusty* target (4 inches of iron on 25 inches of wood).[45] Whitworth's single goal was to penetrate the *Warrior* target before Armstrong, after which he could demand a new competition to determine who would be the cannon king of England. Armstrong's 300-pounder tests on 9 April 1862, indicated that he was close, but, on 25 September 1862, Whitworth became the first man to pierce the *Warrior* target. Moreover, he did so from 600 yards. A second series of tests against a new *Warrior* target, held in early November and witnessed by representatives from the War Office, the Admiralty and a number of civilian engineers, confirmed that the Whitworth shell could penetrate $4^1/_2$ inches of iron.[46]

These two tests were the first shots in a battle between Armstrong and Whitworth over whose gun would best sink ironclads. Armstrong responded immediately to the publicity over Whitworth's success in a letter to *The Times*, a popular forum for technical disputes among inventors. He noted inconsistencies in the data and claimed that the test had been biased in Whitworth's favour. Whitworth supporters within the War Office had conducted it, he said, and it thus represented 'not so much what is to be fairly said upon the subject as what can be said upon the subject by the friends of Mr. Whitworth'. The Armstrong gun, he said, fired 'much slower in the columns of some of the newspapers than they do anywhere else'. Armstrong ended his letter with two important points. Firstly, he claimed that the gun used by Whitworth to penetrate the *Warrior* was in fact made in the Royal Gun Factory on the Armstrong built-up method. Secondly, he proposed a competition to determine whose gun was best suited to sink ironclads.[47]

Whitworth's chance had come and he promptly grasped it.[48] He answered Armstrong's letter a few days later and claimed that the trials in 1858 from which Armstrong had gained so much had been controlled by Armstrong's friends. He pointed out that Andrew Noble, the Secretary of the 1858 Committee on Rifled Cannon, had joined the Elswick Ordnance Company shortly after the competition. Whitworth did not deny that he had used the built-up method to make his recent gun but said he had used that method in 1855. Besides, he added, the general technique of coiled hoops had been known long before Armstrong had come along. Each of them, he said, simply used the idea in their own way. His gun had to be constructed at Woolwich because his Manchester factory was too small. Whitworth admitted that the general manager of the government arsenal, John Anderson, had provided him with assistance in making the gun, but he insisted (rightly) that the gun was rifled in a manner developed by him.[49]

The question of whether or not the gun used by Whitworth was an Armstrong built-up gun was investigated by the Select Committee on Ordnance. The idea of shrinking iron coils around each other had been known for centuries, but Armstrong claimed to have been the first modern industrialist to do it successfully and he was determined to defend that reputation. He meticulously prepared to prove that the gun with which Whitworth had penetrated the *Warrior* was in fact an Armstrong gun.

The crucial witness was John Anderson, a civilian engineer who had been at Woolwich since 1842 and had been appointed Armstrong's assistant in 1859. Armstrong trained Anderson to manufacture the built-up gun and Anderson supervised the building of them in the government factory while Armstrong worked on his 300- and 600-pounders.[50]

The War Office became interested in Whitworth again after his shell nearly penetrated the *Trusty* target in 1860. They requested Anderson to advise Whitworth on the construction of the gun with which he would test his shell against the *Warrior* target. Anderson examined the design of Whitworth's cast-iron gun and, with Whitworth's agreement, changed it into an Armstrong built-up gun of wrought iron. Whitworth's gun was in fact made by Anderson on the Armstrong built-up principle.[51]

Armstrong prepared his case to be heard before the Ordnance Select Committee. He asked Stuart Rendel, the son of his old friend James Rendel, to ensure that the right questions were asked of Anderson and Whitworth by two members of the Committee who supported Armstrong, Henry Jervis and Sir John Hay. Rendel also coached Anderson to ensure he would be firm in his evidence about the construction of the guns.[52] Anderson confirmed that Whitworth had used an Armstrong built-up gun to penetrate the *Warrior* target.[53]

Armstrong also prepared Jervis and Hay to question Whitworth. He asked Rendel to be sure that the correspondence between Whitworth's Manchester Ordnance Company and the Royal Gun Factory, which showed Whitworth had agreed to use a built-up gun, was in front of Jervis and Hay. He sent Rendel to visit Jervis and Hay and acquainted them with the facts of the case about Anderson's role.[54] Rendel also ensured that Jervis and Hay had a firm grasp of the case. Finally,

Armstrong provided specific questions to be put to Whitworth if he repeated his claim which he had made in *The Times* that he had constructed a gun upon Armstrong's coil principle in 1855: he should be asked to produce it and explain why he did not use it at the competition with Armstrong in 1858.[55]

This preparation paid off. Jervis and Hay questioned Whitworth closely and, when forced to state that either Anderson must 'eat the words which he used before the committee' or that his design had been changed, Whitworth admitted the latter.[56] When Whitworth repeated the claim that he had used the coil principle in 1855 and Jervis asked where the gun was, Whitworth claimed he had never finished it.[57]

Armstrong spent a great deal of time and effort to prove his point, but reputation was important in the armaments business and Whitworth had attempted to disprove the originality of the very system which enabled him to obtain his success against the *Warrior* target in November 1862. Armstrong explained to Rendel that he had

> just reason to complain of Mr. Whitworth, Mr. Lyonal Thomas and others having guns made for them in the R.G.F. upon my principles, by my methods and according to drawings supplied by persons who have derived all their information and experience under me – and that these guns should be used in competition with my own guns and should go forth to the world as the inventions of my opponents.[58]

V

Armstrong had successfully defended his reputation as the man who made the first built-up gun, but the fact remained that the *Warrior* target had been penetrated by a Whitworth shell. He was now faced by a competitor who might move ahead of him in armour-piercing shells. Each man proclaimed that he could penetrate the *Warrior* target from at least 1000 yards, the standard upon which the government's forts depended, and each challenged the other to a new battle of the guns.[59]

The competition was organized by the government. Whitworth considered the Ordnance Select Committee committed to Armstrong and demanded that a separate committee conduct the tests,[60] and thus a Special Committee on the Armstrong and Whitworth Guns was formed by the Admiralty and the War Office in January 1863. It was clear that both guns would soon penetrate armour of even greater thickness than the *Warrior*'s and the task of the committee was to compare and contrast the relative merits of the two different kinds of rifling and shells advocated by Armstrong and Whitworth and produce a report. The stakes were very high: the entire fleet would carry only the gun which was deemed better, since the navy could not have guns of such different character in service at the same time. The committee (usually called the Armstrong–Whitworth Committee) was composed of military and naval officers chosen by the government and included a representative chosen by each of the two competitors. Armstrong chose William Pole, an eminent professor of civil engineering who had been an assistant to James M. Rendel in the 1850s and a member of the Iron Plate Committee. However, poor health forced Pole to resign and Armstrong turned to Stuart Rendel for help.[61]

The Rendel family had grown closer to Armstrong since the death of James M. Rendel in 1856. Alexander, the eldest of the four brothers, had taken over his father's engineering practice and played no role in the history of the Elswick company, but two others, George and Hamilton Rendel, also engineers, became partners in the Elswick firm in 1858. Hamilton managed the civil engineering department (the Elswick Engine Works); George Rendel, who had lived and worked with Armstrong for three years after the death of his father, managed the Elswick Ordnance Company.

Stuart Rendel did not follow his brothers into engineering. He studied at Eton and Oxford, trained for the law at Inner Temple, and was called to the bar in 1861. But, as he later recalled, he too was drawn into Armstrong's affairs:

> I had already been useful to [Lord Armstrong] in his official correspondence and in ways for which my university and legal training qualified me, and I became almost necessary to him when he was confronted with serious controversy in the Press and in Parliament.[62]

Rendel felt distinctly handicapped 'as a young civilian, sitting with generals and admirals and in a partisan capacity'.[63] However, he laboured with growing interest over the technological complexities of ordnance, fascinated by the great contest and the endless gun trials on land and sea.[64]

Both guns were constructed on the Armstrong built-up method, but each used different projectiles and employed a distinctive form of rifling. Guns are constructed to accommodate projectiles designed to produce a specific effect on the target. As Armstrong said in 1860, 'the projectile should rule the gun, not the gun the projectile'.[65] The object of the Armstrong–Whitworth Committee was to determine what kind of projectile (and hence what kind of rifling) sank ironclads more efficiently. Armstrong used conical shells with between three and six studs attached to produce the spin. Whitworth preferred flat-headed shells and hexagonal rifling. The Armstrong shell delivered greater explosive power than the Whitworth shell but the latter penetrated deeper. On the other hand, the flat-head shell punched a neat hole in the armour while the conical shell crushed a relatively large area of the side of an ironclad. The flat-head shell displaced an area of iron equivalent to the area of the front of the shell; the conical shell displaced an area of iron far greater than its nose. The Armstrong 300-pounder used 50 pounds of powder; the Whitworth shell weighed only 130 pounds and used 25 pounds of powder.[66] The Armstrong–Whitworth Committee investigated all the permutations and combinations of power, range, accuracy, size of projectile, the amount of charge, muzzle velocity and the damage inflicted on the target by each type of gun.

The committee was formed in January 1863, but Whitworth requested several postponements while he made further improvements to his weapon. Armstrong was ready and protested at the delays.[67] Rendel claimed Whitworth bribed the gunners at Shoeburyness 'to induce greater exertion in favour of Messrs. Whitworth's firm'. Colonel Eardley Wilmot, who had lost his position as superintendent of the Royal Gun Factory when Armstrong took over in 1859, and was now active on behalf of Whitworth, challenged Rendel to prove the charge.[68] Whitworth declared that Armstrong guns were made from steel which had come from Whitworth's company. His boasts that Elswick was paying him a royalty disturbed Armstrong enough to investigate.[69] Rendel's friends inside the Royal Arsenal kept him informed of Whitworth's progress.[70] These petty aspects gave a bitter and personal edge to the real question of whether enemy ironclads should be hit with flat-head or conical shells, whether the object was to drill a neat hole through an ironclad or to crush its side. Each side fought hard to have its answer to this question prevail.[71] The military and naval members of the committee were left to make the decision.

When the competition began, Whitworth could claim to have accomplished the primary goal of penetrating 4½ inches of iron, but in December 1863, Armstrong's 600-pounder 'monster' gun was tested at Shoeburyness. This gun was of unprecedented size: it measured over 15 feet long, weighed 22 tons, had a diameter of over 4 feet and a bore of 13½ inches.[72] It threw its 600-pound shell 4½ miles.[73] The shell completely penetrated a floating *Warrior* target from 1000 yards[74] in a test which strengthened Armstrong's case for explosive power over simple penetration.

On 26 December 1863, *The Illustrated London News* published descriptions and illustrations of the damage inflicted on the *Warrior* target. The illustration of the back of the target showed substantial damage and was 'right enough' for Armstrong, but the illustration of the front showed a hole not much larger than one made by a Whitworth flat-head shell and 'a very inadequate idea of the injury [was] in consequence conveyed'. Armstrong told Rendel:

> Now it is very important that the Public should know to the full the damage that was done and they will be much more impressed by pictures than descriptions ... Under colour therefore of supplying *additional* information (not of correcting errors) I want to get the enclosed communications and sketches inserted in the next number of the Illustrated paper.[75]

Rendel persuaded the editor to publish a second illustration of the front of the target which conveyed a stronger sensation.[76]

In March 1864 it was finally and clearly demonstrated that guns could break armour – and that gun makers had overtaken ironmasters – when Armstrong's 600-pounder fired a 345-pound steel shell which broke through John Brown's 11-inch iron plate backed by a 6-inch plate made by Sir William Fairbairn. The shell struck the front target at 1560 feet per second. Never, said *The Times*, had a greater blow been struck by a human agent:

> The mass of steel driven by the tremendous charge of powder must have struck the target with a power almost inconceivable, for everything went down before it. The solid oak beams behind the plate itself hurled bodily back against the Fairbairn target and split into two pieces. ... The 11-inch plate ... was torn apart.[77]

It was a moment of triumph for Armstrong, the culmination of ten years of work to prove the principle of the built-up gun. It did not, however, end the competition with Whitworth, which continued for another year, with performance data piling up. The problem of finding an objective measure remained. As the Duke of Cambridge, who had attended the opening trial, recorded in his diary, all the guns tried were 'admirable in their way, and leave but little to choose between them'.[78] *The Times* too reported that the results were still inconclusive.[79] There were still Whitworth supporters and Armstrong supporters in the War Office and on the special committee, each of whom could point to some evidence from the two years of trials to support their convictions. Herein lies the dilemma of establishing new technological standards. Technological superiority can only be established according to specific contexts in which advantages and disadvantages intermix and overlay each other. Superior performance needs to be defined, as does the meaning of test results. The complexities of the new armaments and the changes they were bringing about in ships left room for doubt and uncertainty as to the future direction British ordnance development should take. The inconclusive results gave added importance to the final stage of the inquiry, the writing of the report. If the question which had been put to the committee in 1863 could not be answered on the testing range, it would be answered in the committee rooms where non-technological factors took precedence.

The excitement of these events remained with Rendel throughout his life. His description of them 40 years later clearly shows the political nature of determining technological superiority and standards, in what is today called the 'social construction' of technology:

> When ... we came to the report stage and for weeks and months together debated our report, I found myself overstrained. Practically I led the case for Armstrong, while General Simmonds ... led the case for Whitworth, and how we came to produce a report fairly capable of construction as generally favourable to Armstrong I can scarcely understand. But so it was, and I may here declare unblushingly the plain fact that but for me the Whitworth case would have triumphed.[80]

However, the Armstrong built-up method of constructing large naval guns was never in question and, after 1865, the conical shell and shunt rifling were confirmed as the standard for British ordnance. Whitworth, beaten again, continued to manufacture and sell his guns to a few countries in the world, but he never achieved a place among the world's leading armaments manufacturers. He died in 1887, and Armstrong bought his company in 1896 in order to obtain its armour plate facilities.

The size and power of guns had increased enormously over a ten-year period, with the greatest leaps taking place within the last five years. In 1855, the most powerful gun in the British arsenal was the cast-iron 68-pounder which weighed less than 5 tons; in 1865, the most powerful gun was the Armstrong 600-pounder which weighed 22 tons. In 1860, an armour plate of only two inches was sufficient to stop any shot or shell; in 1865, an 11-inch armour plate was smashed to pieces. Moreover, the old assumption that guns of such mighty power would be too heavy and produce too much recoil energy for any ship to absorb safely was proved wrong. Large guns could be fired from turret ships, as Cowper Coles had argued and as Ericksson's *Monitor* had demonstrated. That was the true significance of the inconclusive battle at Hampton Roads, and one that did not escape Armstrong. He realized the world was on the brink of a new era in naval power based on the ever-increasing size of ordnance. On the day of his first success against the *Warrior* target he wrote to *The Times* of the future:

> But we must be prepared for vessels stronger than these – stronger even than the Warrior, and should, therefore, go on increasing the size of our guns until we reach some practical limit. The weight of such guns, however objectionable, must be accepted as a necessity, and ships must be adapted for their reception.[81]

Whatever the technological difficulties, he said, the British people could

Note: Note that the built-up construction is identical to that of the 6-pounder in Figure 1.1.

Source: *Illustrated London News*, 26 December 1863, p. 645.

Figure 3.1 The Armstrong 600-pounder 'monster' gun

Note: This was the largest 'built-up' gun.

Source: *Illustrated London News*, 29 July 1876, p. 117.

Figure 3.2 Shipment of the 100-ton gun at the Elswick Ironworks

Note: The small, neat hole was too similar to that made by the Whitworth shell.

Source: *Illustrated London News*, 26 December 1863, p. 645.

Figure 3.3 Effects of the Armstrong 600-pounder on the Warrior *target, front view*

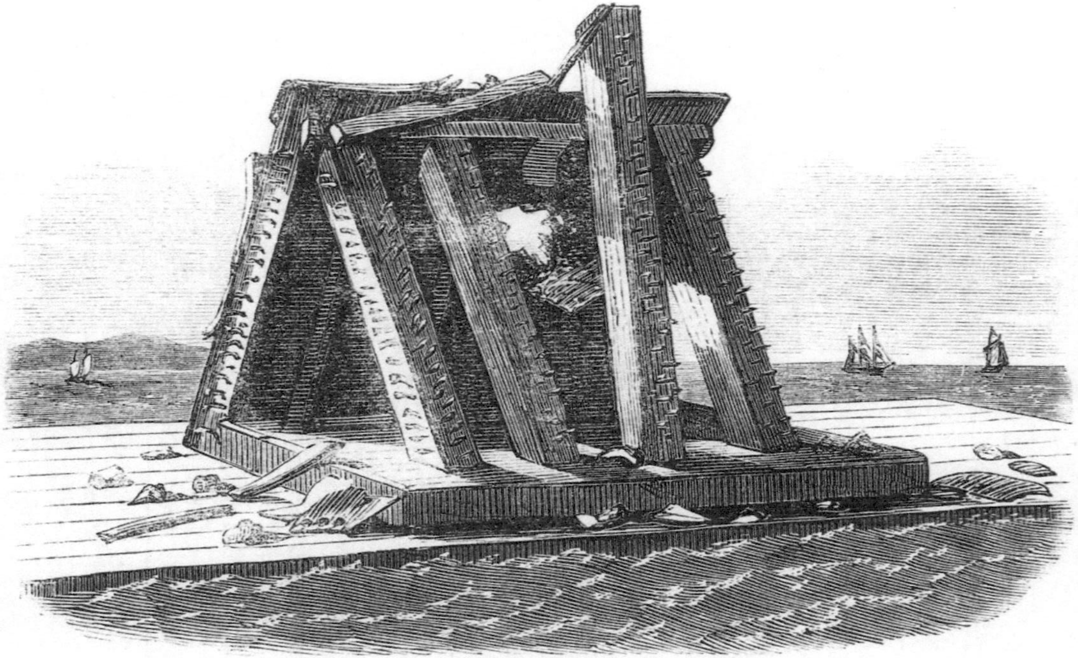

Source: *Illustrated London News*, 26 December 1863, p. 648.

Figure 3.4 Effects of the Armstrong 600-pounder on the Warrior *target, rear view*

Source: *Illustrated London News*, 9 January 1864, p. 37: 'In our Number of the 26th ult. we gave Illustrations of Sir Wm. Armstrong's 600-pounder gun and the floating target which had been built to represent the side of the Warrior. Since the experiment to which these Illustrations referred the target has undergone a most careful survey; and, pursuing the subject we are now able to give additional particulars of the effect produced by the penetration and explosion of a single shell.'

Figure 3.5 The 'corrected' effects of the 600-pounder on the Warrior *target*

Source: *Illustrated London News*, 20 August 1864, p. 181.

Figure 3.6 Inspecting La Gloire *target during artillery experiments at Shoeburyness*

Source: *Illustrated London News*, 20 August 1864, p. 189.

Figure 3.7 Shoeburyness artillery range

'rest assured that they [were] of a nature to be overcome by engineering skill'.[82]

The era of the great muzzle-loaders began in 1865 and reached its climax in 1874, when Armstrong manufactured 100-ton, 18-inch guns for the Italian navy's new turret ships, *Duilo* and *Dandolo*. The British responded in kind with the *Inflexible*, which carried 22 inches of armour, but by 1880 a 16-inch gun could send its 1800-pound shell through 34 inches of wrought iron from 1000 yards. After 1880, steel plates battled it out with the new breech-loaded ordnance in an era of accelerating technological prowess. The 11-inch steel armour on the *Dreadnought* was far tougher than the 22-inch of iron on the sides of the *Inflexible*, and the former's 12-inch guns were incomparably more deadly than the latter's 16-inch guns. The number of heavy guns on warships fell sharply: from the 100 or more on the old line-of-battle sailing ships to 36 on the *Warrior*, to only four on board the *Duilo*, the *Dandolo* and the *Inflexible*. The *Dreadnought* carried 10 guns of unprecedented power.

Throughout this period ships adapted themselves to accommodate the new ordnance as they had always done, a process that involved much controversy among naval architects. The first generation of ironclads – the *Warrior* class – were traditional broadside ships in which their guns were arranged in a row along the sides under the deck and protected by a strip of 4½ -inch iron (the last of this class, the *Northumberland*, carried Armstrong 12-ton muzzle-loaders – the 300-pounders – and 5½ inches of armour from end to end), but the weight of the 6-inch, 9-inch and 12-inch armour necessary to stop the new ordnance added enormous weight to the ships (far more than the weight of the new guns) and hence only a few guns could be fully protected by the thickest armour. Ships increased in size to carry the heavy plate: the *Warrior* displaced 6000 tons, the *Duilo* and *Dandolo* 11 000 tons, the *Dreadnought* 18 000 tons. Similarly, armour accounted for most of the rising price of battleships: without guns, a nineteenth-century line-of-battle sailing ship had cost £115 000; the *Warrior* cost £350 000, the *Duilo* £700 000, the *Italia* (1886) £1 million; the first *Dreadnought* (1905) cost £1 700 000. The armour amounted to as much as 70 per cent of these totals.[83]

Technological developments in large naval guns accelerated and those responsible for Britain's navy were thrown into a state of uncertainty. The First Lord of the Admiralty, the Duke of Somerset, expressed the confusion which confounded the Admiralty in the early 1860s:

> One day we are told the ships must carry larger guns. Another day we are told they must go very fast. If they are to carry great weight and go very fast, they must be enormous vessels, the only way to

> carry heavy armaments is to give up somewhat of the speed we have endeavoured to obtain.[84]

Disputes broke out among naval architects over the choices that had to be made and the balances struck between heavier guns and thicker armour (with Armstrong naturally a strong advocate of the former), weight and size of ship, and speed, coal capacity and the area of sail (a question related to the range desired by Britain's strategic planners and of course dependent upon coaling stations throughout the world). As new guns came on the market later in the century new ships were developed both to carry them and to defend against them.[85] Torpedo boats were designed to carry torpedoes, a new kind of projectile developed in the second half of the nineteenth century, which was answered by the quick-firing gun carried in the torpedo-boat destroyer (later called simply the destroyer).

Two battleship designs emerged in the 1860s in response to the new ordnance, each with its supporters and detractors. Some, led by Cowper Phipps Coles, called for turret ships, made popular by the *Monitor*, which carried guns in one or two rotating turrets located in the centre of the deck; others, led by Sir Edward Reed, the Admiralty's naval constructor, preferred the central battery ship, which retained much of the design of the broadside but concentrated its guns and its thickest armour in a central battery. The turret ship was low in the water (a low freeboard) and thus could not venture safely out to sea; the central battery design maintained a high freeboard for high seas duty. Turret ships could carry the largest guns and therefore Armstrong supported the building of more of them. That they happened to be suitable only for coastal defence, of which he was a strong supporter, was secondary. When sea-going ships could carry the most powerful guns in the 1880s, he advocated the protection of British trade routes. The dispute between those who advocated turrets and those who called for central battery ships hinged on the question of sails, sea-worthiness and coal capacity, questions that had not been completely resolved since the introduction of steam and iron. The sinking of the badly-designed turret ship *Captain* in 1870 did not settle the matter in favour of the central battery. Both were built during the period of transition between 1860 and 1885.[86]

VI

When the 'monster' muzzle-loaders entered their days of glory in the 1860s, the Armstrong breech-loaders, so enthusiastically embraced in 1858–9, lost their place in Britain's military and naval arsenals, and with them went the Elswick Ordnance Company. In 1860, Armstrong's breech-loaders underwent their first test in battle. China, which became one of Elswick's best customers in the 1870s, was the first country to come under fire from Armstrong breech-loaders. Relations between Britain and China had settled down after the opium wars of the 1830s and 1840s, but a new period of hostility was touched off in October 1856, when Chinese officers boarded the lorcha *Arrow* as it lay at anchor in Canton harbour and arrested 12 of her Chinese crew on charges of smuggling (her British captain was not arrested). The vessel, owned by a Chinese resident of Hong Kong, had been authorised to sail under the British flag by a Colonial Ordinance. Although the registration had expired, the British government insisted that the Chinese actions violated a Treaty of 1843 which required China to make such arrests through British officials. By invoking treaty rights, the British government ensured that the incident rapidly escalated into a diplomatic crisis. An ultimatum was issued for the Chinese publicly to return all 12 men, apologise and promise to respect the British flag in the future or face British naval action.

Chinese authorities denied the *Arrow* was a British ship and claimed that she was not flying the British flag when the officers boarded her. British gunboat diplomacy was launched: Canton was bombarded and occupied in 1857. The Chinese Imperial High Commissioner commanded the people of Canton and Hong Kong to rise up against the 'English barbarians'. The Taku forts which guarded the Peiho River were taken in 1858, opening the way to Tientsin and Peking. Negotiations led to the Treaty of Tientsin, which opened more Chinese ports to foreign trade and established the Chinese Maritime Customs Service under British supervision (an act which later had a considerable effect on the growth of Armstrong's company).[87] China failed to ratify the Treaty in 1859 and British warships again attacked the Taku forts (which had been returned to the Chinese). However, a British gunboat expedition up the Peiho was badly mauled and a larger Anglo-French force was organised and sent out in 1860. This expedition carried Armstrong breech-loaders.[88]

Palmerston was reluctant to send the new guns on this joint venture because he did not want to give the French an opportunity to examine them, which had been the primary reason for the Derby government's eagerness to obtain the patent rights to the guns. Palmerston had the same concern when he became prime minister in June 1859 and suggested to Herbert in October that smooth bores would 'probably do well enough for the Chinese'.[89] However, since the opportunity to observe the guns in action was also important, some of them were sent to China as supplements to the standard muzzle-loaded artillery. The commander of the expedition, Sir James Hope Grant, praised the accuracy of the guns and the power of the Armstrong shell.[90] The Duke of Cambridge believed that 'the efficiency of the Armstrong guns [was] completely established' and considered that their successful performance was 'even more important than the whole expedition to China'.[91] *The Times* called the field guns 'in every respect a success': range, accuracy, power, ease of firing, and the small amount of fouling which required only infrequent washing. The breech mechanisms, which many considered too fragile, had withstood the long and rough journey to China. The Armstrong guns cost more than muzzle-loaders, said the editor, but their performance demonstrated that they were 'a good investment' and agreed with the soldiers in the field who claimed that 'the British army had got such a weapon as no army ever possessed before'. *The Times* recommended breech-loaders for all branches of British defence forces: the army, the volunteers, in forts and on board ships. Armstrong's guns were 'not likely to be practically superseded for some time to come'.[92]

These endorsements soon had their effect in the factories at Woolwich and Elswick. Orders to Elswick, which had amounted to under £8000 in 1858–9 and £30000 in 1859–60, leapt to nearly £400000 in 1860–61 and remained at that level in 1861–2. About 3000 Armstrong guns of all sizes were manufactured, of which half came from Elswick. The 12- to 40-pounders went to the army as field pieces, the 70- and 110-pounders to the navy.[93]

The 110-pounder did not enjoy the same success as the smaller field guns. There were two complaints against it. Firstly, it did not penetrate iron. This had been confirmed in tests against the *Trusty* target (4 inches of iron) in 1859 and against the *Warrior* target in 1860.[94] The results of these tests convinced Armstrong that breech-loading had reached the limits of its strength with the 110-pounder and he turned his attention to building muzzle-loaders.

But the 110-pounders put aboard the *Warrior* and many other British warships were never intended to sink ironclads like *La Gloire*. The old 68-pounder smooth-bore muzzle-loader was more powerful than the 110-pounder at close range (200 yards and less) and was designated to confront enemy ironclads until the 300- and 600-pounder Armstrong muzzle-loaders were installed in the forts. The 110-pounder rifled gun could not fire its much heavier shell with as high a muzzle velocity as the 68-pounder fired its shot but, as was indicated earlier, round shot lost velocity more rapidly than elongated shell. Beyond 200 yards the Armstrong 110-pound shell had greater momentum than the 68-pound shot.[95] The 110-pounder could not sink an ironclad but its accuracy and range would inflict great damage and kill many sailors from beyond the range of the enemy's guns. It was the best gun for use against wooden vessels, which still constituted the vast majority of the world's warships. It was also an effective gun in amphibious operations.

These points were forgotten in the parliamentary debate over guns, forts and armour. Fierce attacks were launched against the Palmerston government for deploying Armstrong 110-pounders which could not penetrate armour at even the closest range. Ralph Bernal Osborne was one of the most vigorous critics. Osborne, the eldest son of a prominent Whig, had been secretary of the Admiralty in the Aberdeen government and during the first Palmerston government. He was an effective and entertaining speaker but, since the First Lords of the Admiralty, Sir James Graham and then Sir Charles Wood, also sat in the House of Commons, Osborne had little chance to perform. He was left out of Palmerston's second government. In 1861, Sir George Lewis, who sat in the House of Lords, became Secretary of State for War.

The undersecretary, Earl de Grey, was also a peer and it was decided to replace him with a man who sat in the House of Commons. Palmerston suggested Osborne because he was an excellent speaker but, since Osborne had once criticised the Horse Guards, the Duke of Cambridge, who in fact liked Osborne, thought the appointment would create disharmony between the Horse Guards and the Secretary of State for War. The position therefore went to Baring, whom the duke thought 'a very right thinking and safe man'.[96]

A disgruntled Osborne joined the growing chorus against the forts, Elswick's monopoly and Armstrong's position inside Woolwich. He launched a blistering attack against the government for misrepresenting the results of the tests of the Armstrong gun carried out on 9 April 1862. Despite £3 million spent on Armstrong guns, he said, the best gun for use against ironclads remained the old 68-pounder smooth bore.[97] Neither Lewis nor Sir John Hay pointed out that the 9 April test had been with an Armstrong 300-pounder muzzle-loader and that the £3 million had been the amount spent on all sizes of Armstrong breech-loaders for both the army and the navy and that the smaller ones had proved excellent field pieces. The government also failed to note that the 110-pounder had not been expected to sink ironclads and that neither the government nor Armstrong had ever pretended that it would.[98]

About this time, however, the second and more serious complaint was made against the 110-pounder. Charges arose that its breech mechanism often burst when the gun was fired, injuring gunners. Lewis would not release a report, but he claimed,

> the evidence which has been produced convinces me that the statements with respect to the failure of the Armstrong guns are either entirely erroneous or grossly exaggerated, and that, upon the whole with the exception of certain slight defects which have been

> remedied by subsequent improvements, the guns of Sir William Armstrong have answered all the purposes that were expected of them.[99]

Sir John Hay insisted that 'the Armstrong gun was as great an improvement on the smooth-bore 68-pounder as the Enfield rifle was upon the old Brown Bess'. It was only 'exceptional circumstances' which had 'occasionally caused some failures, which had been magnified'.[100] The breech mechanism on the 110-pounder did need strengthening. In the smaller guns used by the army the vent piece was strong enough to withstand the explosion of the charge but the more powerful explosion in the 110-pounder approached the resistance limit of the larger vent. After a number of shots, the vents could be damaged or blown out altogether. Rumours of blown vent pieces, which Armstrong tried to explain, persisted in 1860–61.[101]

Armstrong had sat on a committee in September 1859 which had determined the specifications of the 40-pounder (which was soon built up into the 70-pounder) and that gun performed without accident. However, in 1858–9, the navy, suffering from its loss of reputation as the defender of British shores, wanted the largest breech-loader it could obtain for the *Warrior*. In October it ordered 110-pounders be placed aboard its ships before trials could be conducted. As the Ordnance Select Committee later explained:

> The political necessities of the day appear to have been so urgent as not to allow time for maturing the design previous to its manufacture. In consequence of the excessive pressure which existed at that time for the supply of guns of that calibre, the first 100 of these guns were constructed before any experiments upon them had been concluded.[102]

Armstrong field guns had been carefully tested at Shoeburyness in 1858 and proved themselves in battle in China in 1860, but the 110-pounder had yet to be tried in battle. The opportunity came in Japan.

As in China, the intrusion of foreigners into Japan created cultural and political tensions which occasionally resulted in incidents of violence. In September 1862, near the town of Kagosima, a British subject, Charles Richardson, gave offence to a local political authority as he passed by. Members of the procession confronted Richardson and his English companions and an assault ensued in which Richardson was killed and his English companions severely wounded. The local political authority, the Prince of Satsuma, refused to dispense justice and his privileges and immunities shielded him on this matter from pressure from the central Japanese government, which had apologised and paid compensation. The crime remained unresolved and, in August 1863, the British government demanded that the Prince of Satsuma immediately try and execute Richardson's murderer and pay an indemnity to his family and to those who had also been assaulted at the procession or face reprisals. The prince temporised and on 14 August a British squadron under the command of Vice-Admiral Augustus Kuper confiscated three of the prince's steamers (built in England) which were anchored in Kagosima Harbour. As Kuper towed the ships out of the harbour, the Japanese opened fire from their forts and a three-day battle took place. One fort was destroyed but a fierce storm blew up and some of the shells from the rolling ships landed in the town of Kagosima, setting its wooden buildings on fire (for which Kuper was criticised in Parliament).[103]

Armstrong 40- and 110-pounders were used in this exercise in gunboat diplomacy. The 40-pounders performed without trouble but vents blew out on the 110-pounders and triggered more attacks on their efficacy during March and April 1864. Ralph Bernal Osborne pronounced the Armstrong breech-loaders a 'dead failure' and demanded that the government consider the guns made by Horsfall, Whitworth and Blakely.[104] Rear Admiral Lord Clarence Paget, the secretary to the Admiralty, shifted the responsibility for the faulty guns onto the War Office. The Admiralty did not manufacture its guns he said (correctly), and claimed (incorrectly) that the navy had no control over the quality of guns they received.[105] General Peel, who had been the Secretary of State for War in the Derby government, defended the War Office. The Admiralty had ordered 500 of the 110-pounders in 1859, he said, despite Derby's caution to await trials of guns larger than the 40-pounders. However, Peel understood the 'great pressure' upon the navy at the time.[106] On the other hand, the Admiralty had known about the weak vent pieces for a year but had done little to replace them with one of a different design.[107]

Armstrong reacted swiftly upon reading Paget's speech in *The Times*. He wrote to Paget to say that, although the 110-pounder needed improvements, the publicity about its poor performance in Japan was exaggerated. He reminded Paget that he, Armstrong, had told the Admiralty in 1859 that the vent piece in the 110-pounder was inadequate, but his caution was brushed aside in 'the haste and pressure' of the moment, and he protested that now 'authoritative sources' were leading the public to underestimate a valuable gun and that it was unfair to connect the Armstrong name 'so emphatically' with the least successful of his breech-loaders while at the same time failing to emphasise the excellent results obtained by the Armstrong 6- and 12-ton muzzle-loaders. Armstrong was aware of the importance of public opinion in these matters and published his correspondence with Paget in *The Times*.[108]

Armstrong expected that Paget would seek 'revenge' by 'raking up every case of broken vent pieces which has occurred to 110-pdrs' and he asked Stuart Rendel to induce Henry Jervis, who had supported Armstrong on the Ordnance Select Committee, 'to take up the case'. The gun had performed without accident in recent tests and Jervis, said Armstrong, should attribute the accidents at Kagosima to the improper or inadequate training of the gunners.[109] The near-hurricane storm and the fact that the British ships were inside a harbour and under close fire from shore batteries meant that the training of gunners became a crucial factor in the performance of the 110-pounders. Under the extraordinary pressures of battle at Kagosima, the British gunners may have hurried too much and failed to tighten the breech screw firmly.[110]

Five months later, the 110-pounder was used again in Japan. The anti-foreign movement escalated during the summer of 1864 and another allied expedition was sent out. For four days in early September, British, French, American and Dutch warships pounded the Japanese forts at Simonosaki. The Admiralty ordered a full report on the performance of the Armstrong breech-loaders from the admiral in charge of the operation. Captains of the British ships reported that, while the 40-pounders had fired without accident, 110-pounder vent pieces had blown out, and the admiral concluded that the gun was not suitable as a naval weapon.[111] This recommendation, coming just as the Armstrong–Whitworth Committee was writing its report in favour of rifled muzzle-loaders, sealed the fate of the 110-pounder and it was withdrawn from service. In fact, in 1865, *all* breech-loaders were withdrawn from both the navy and the army. The debate over the relative merits of breech-loading and muzzle-loading continued,[112] but breech-loaders did not return to the British arsenals until the 1880s.

The apparently retrograde decision to return to muzzle loading for even small artillery and ordnance pieces, whose breech systems had proved satisfactory, was based on financial and technical calculations. In 1862, a young Irish major, William Palliser, patented a method by which cast iron, smooth-bored artillery and naval ordnance could be rifled. For ten years, Palliser had been experimenting with rifling guns by inserting rifled wrought iron tubes into the barrels of old smooth-bored muzzle-loaders, and transforming them into rifled muzzle-loaders. Palliser's innovation meant that the very large inventory of cast iron guns held by the War Office could now be converted into rifled guns at one-third the cost of manufacturing new rifled muzzle-loaders.[113] The Duke of Cambridge, who had been one of the first and most enthusiastic supporters of Armstrong breech loaders, demanded proof that the advantages of converted muzzle-loaders were enough to warrant the very high costs of changing back.[114] The technological value of Palliser guns was proved on the firing range and, since converting a gun to rifled muzzle-loader cost far less to carry out and much less to maintain than breech-loaders, the War Office purchased the rights to use Palliser's conversion techniques and stopped all production of breech-loaders; Woolwich began converting cast-iron smooth bores into rifled muzzle-loaders for the British army and navy.[115]

General John Adye reopened the question of breech-loading field guns when he became Director of Artillery in 1870, but his confidential report concluded that the Armstrong breech-loaded gun of ten years earlier had not fulfilled its promise. He was, he said,

> bound to point out the grave defects which have since become apparent in the Armstrong system, and I am further compelled to state that these defects began to develop themselves from the very commencement; and some of them become aggravated rather than lessened by time and experience.[116]

Adye listed the persistent problems associated with the Armstrong breech gun: the 'complicated' and 'delicate' breech system; the sophisticated ammunition; the need for constant attention; the necessity to use the best and most costly materials; the fact that 'skilled artificers and special tools' were required to service and repair it; the need for trained gunners to operate it without accident or breakdown. Armstrong had always insisted that the difficulties with his breech-loaders were mostly due to untrained or careless gunners, or the poor manufacturing with inferior materials carried out at Woolwich. However, Adye's report stated that rifled muzzle-loaders were also superior in range and accuracy to breech-loaders and that their simpler, sturdier construction made them 'better adapted for the rough purposes of war'. Finally, Adye justified the retention of muzzle-loaders on the grounds that they were much cheaper to build and maintain than breech-loaders,[117] an important consideration in the era of Gladstonian 'retrenchment' in armaments expenditures. Each type of gun had a different firing drill, which could create difficulties during battle if gun crews had to switch to a different system. It was rifled muzzle-loaders that had penetrated iron plate and restored the British Navy to its traditional place as the protector of England, and breech-loaders had little tactical role in the ironclad era.

The international situation had also changed dramatically since 1859. British industrial might had produced an ironclad fleet far superior to the French and Napoleon III turned to the east for a foreign policy. The threat of a Franco-Russian alliance was eliminated in 1863, when Napoleon lent rhetorical support to the Polish insurrection against their Russian masters. The British strategic position was further strengthened as Franco-German relations deteriorated throughout the decade. In the 1870s, the main objective of the new French Republic was to recover from its by Prussia in 1870. The Russian navy was important only in alliance with the French and this possibility did not arise again until the 1880s. There was no German navy at all, and the once-feared rise of a global American navy was eliminated by the Civil War and the domestic concerns with reconstruction and industrialisation which followed. It was not until the 1880s that the United States was able to embark on the building of a new navy.[118]

VII

Breech-loaders were officially withdrawn in 1865, but the government had stopped buying them from Elswick in 1862. The panic of 1859 had passed and demand for them could be filled by either Woolwich or Elswick; the government abandoned Elswick in September 1862.[119]

The Elswick Ordnance Company had arisen fast and high and it fell fast and hard. In 1859, the Armstrong breech-loader had the unequivocal support of virtually all political, military and naval leaders in Britain. Woolwich smooth-bore muzzle-loaders were pronounced inadequate and the Armstrong gun was to be introduced immediately. Resistance was not tolerated: the superintendent of the Royal Gun Factory, Colonel Eardley Wilmot, did not support the Armstrong gun with sufficient enthusiasm and was therefore fired. His job went to Armstrong. Colonel E.M. Boxer, inventor of the famous Boxer shell, and for whom the ammunition shop at Woolwich had been built in 1854, was instructed to produce only Armstrong shells. Sidney Herbert explained the new situation in a letter to Sir James Graham:

> It is indeed a time of great care and anxiety, I am preparing a paper for the Cabinet on the amount of force to be maintained. It will be a heavy bill. I do not know how Gladstone will stand the demands made on him by Seymour and myself. I am pressing on arms, cannon, ammunition works: everything is going on at full work. I have closed altogether the Royal Gun Factory at Woolwich, not without much gnashing of teeth. But I satisfied myself that the concern was a failure, not producing better guns or cheaper guns than the French, and, what is worse, not producing them at all. The whole building will now be thrown into the Rifle Cannon Factory under Armstrong, and will make his guns, and hoop cast-iron guns, to strengthen them for rifling.[120]

The Elswick Ordnance Company was the only private company allowed to manufacture Armstrong breech-loaders and its fortunes rose and fell on those guns. In 1858–9, the company sold guns worth under £8000 to the government; in 1859–60 it sold over £30 000 of guns; in 1860–61 and 1861–2, revenues were £400 000. Total orders for breech-loaders (that is, to both Elswick and Woolwich) in 1863–4 amounted to £285 895, 70 per cent of what Elswick alone had produced in 1860–61 and 1861–2. The company's share of this total came to just over £123 000. In 1863–4, Elswick manufactured slightly over £6000 of breech-loaders for the government. Total sales in the five-year period 1858–63 were just under £1 million, of which nearly £800 000 came in the two years 1860–62.[121] Under the terms of the 1859 agreement, the company was not allowed to sell to anyone else without the consent of the British government, and this was refused. Even George Rendel's desperate request to sell a few guns to merchants whose vessels sailed pirate waters was turned down. By December 1862, the company was virtually bankrupt and entered a claim for £137 000 compensation, proposing that the government buy the company's assets for that amount. This too was refused. On 28 January 1863, the government formally terminated the contract and Elswick submitted another claim, for £118 000, and, following several months of valuation investigations by referees, settled for £85 000.[122] The government bought all the moveable machinery.

By early 1863, it seemed there was no hope. Production had stopped and the company was winding up its affairs. Armstrong had sold the manufacturing plant to the government and was willing simply to carry on his work on the monster muzzle-loaders at Woolwich (he was still the Engineer of Rifled Ordnance). Armstrong himself was a wealthy man and socially eminent. He was 53 years old and had just purchased a large estate in Northumberland to which he might retire once he had developed his rifled muzzle-loaders. He was inclined to accept gracefully the demise of the Elswick Ordnance Company.

His partners and associates were not content to see their investment and years of work slip away without a struggle. Stuart Rendel was in the midst of promoting the Armstrong muzzle-loaders on the Armstrong–Whitworth Committee and argued that, if the British government were unwilling to buy them from Elswick, many other governments were eager to purchase the famous Armstrong guns. Armstrong considered foreign arms sales unpatriotic, but Rendel insisted that foreign sales benefited England, since other nations would come to depend upon her for their weapons. Armstrong was not persuaded. In the end, it was his sense of obligation that changed his mind. Andrew Noble had given up his career at the War Office at Armstrong's urging, to join Elswick, and the other men at Elswick, whose lives had been changed by Sir William Armstrong, pressed him not to let the company die. Armstrong said

> The friends who compose that company, and who were induced through me to embark their capital in the undertaking, feel that they have a claim upon me to join their firm, and lend my aid in rendering their uncovered expenditure productive.[123]

Accordingly, on 5 February 1863, Sir William Armstrong resigned from the Royal Gun Factory and returned to Newcastle.[124] Large guaranteed orders for naval guns and the use of Woolwich for research and development were luxuries of the past.[125] On the other hand, the Armstrong gun was famous among military and naval experts throughout the world – a reputation soon strengthened by his success against the *Warrior* in December 1863 and the breaking of the 11-inch plate in March 1864 – and, although Britain was entering an era of peace and retrenchment, other nations were less fortunate. Indeed, in the 1860s, there were plenty of wars and threats of war to ensure an expanding global market for armaments. The Elswick Ordnance Company was established to supply that demand and sell the latest weapons to any government.

Notes

1. Lord Stanmore, *Sidney Herbert of Lea: A Memoir*, vol. 2 (London: Murray, 1906), pp. 211–12.
2. Ibid., p. 212.
3. The document contains 92 pages, of which 30 are irrelevant reports and memos dating back to 1825. Only 1029 questions were asked of less than a dozen witnesses. Lord Overtone contributed a two-page report on how disruptive a French invasion would be to English trade and commerce, presumably attached to alleviate any middle-class reservations.
4. 'Report of the Royal Commission on the Defence of the United Kingdom', *Parl. Papers* XXIII (1860), pp. 432–3.
5. Ibid., pp. x, xii.
6. 'Copy of Minutes by the Defence Committee, Relative to the Report of the Royal Commissions for National Defence', *Parl. Papers* XLI (1860), p. 621.
7. Michael Partridge, 'Military Planning for the Defence of the United Kingdom, 1814–1870', PhD. dissertation, London School of Economics, 1984, pp. 264–5.
8. Michael J. Salevouris, *Riflemen Form: The War Scare of 1859–1860 in England* (New York: Garland, 1982), p. 252. Partridge, 'Military Planning', p. 256, suggests that Key's support for increasing the army budget can only be explained by his becoming 'imbued with the prevailing ideas' of the moment. Perhaps, but Key did not suggest that the British navy was unable to stop an invasion, only that its dockyards should have extra protection. The important point is that, whatever Key's reason for supporting the forts, it got him onto the Commission.
9. *Dictionary of National Biography* (*D.N.B*).
10. Ibid.
11. Ibid.
12. Ibid. (Supplement, vol. 22), p. 914.
13. 'Report on the Defence of the United Kingdom', *Parl. Papers* XXIII (1860), qs 863–4, 898, 901, 937–8.

14. *Hansard*, 23 June 1862, col. 885.
15. 'Great Britain', *Parliamentary Papers* XXXVI (1860), pp. 495–511. Partridge, 'Military Planning', p. 267 claims that it was doubts about the forts raised by Armstrong that were crucial: 'To quieten him the War Office reassembled the Royal Commission.' Armstrong agreed with Coles that the guns could be carried on ships. We shall return to this point later. On Coles, see Stanley Sandler, *The Emergence of the Modern Capital Ship* (Network, Delaware: University of Delaware Press, 1979).
16. The best description of this famous encounter is provided in Richard Humble (ed.), *Naval Warfare: An Illustrated History* (London: Orbis Publishing Limited, 1983), pp. 162–4. See also Antony Preston, *History of the Royal Navy* (London: Hamlyn, 1983), pp. 19–20. For an analysis of the significance of the encounter, see Bernard Brodie, *Sea Power in the Machine Age*, 2nd edn (Princeton, N.J.: Princeton University Press, 1943), pp. 171–2, 182. Brodie argues (pp. 206–7) that the battle at Hampton Roads and two other naval encounters during the American Civil War in fact revealed that ironclads were no longer invincible.
17. *Hansard*, 31 March 1862, col. 263.
18. The Spithead forts had an estimated cost of £1 million, but Sir Frederick thought that £2 million would be closer to the eventual price, excluding the cost of the 380 (Armstrong) guns and the 2700 gunners. Ibid., cols 264–6.
19. Ibid.
20. Ibid., cols 269–72.
21. Ibid., cols 272–4.
22. Ibid., col. 275.
23. Ibid., col. 285.
24. Sir J. Emerson Tennent, *The Story of the Guns* (Richmond, Surrey: The Richmond Publishing Co. Ltd., 1972), pp. 234–5.
25. *Hansard*, 14 June 1861, col. 1058.
26. *Hansard*, 4 April 1862, cols 580–630.
27. F.S. Stoney and C.A. Jones, *A Text-Book of the Construction and Manufacture of the Rifled Ordnance in the British Service*, 2nd edn, (London: H.M.S.O., 1872), pp. 2, 14–17.
28. *The Times*, 10 April 1862, p. 9.
29. *Hansard*, 13 June 1862, cols 530–32.
30. *Hansard*, 23 May 1862, col. 2151.
31. *The Times*, 10 April 1862, p. 9.
32. *Hansard*, 11 April 1862, col. 842.
33. Ibid.
34. Ibid., col. 846.
35. Palmerston to Victoria, 22 June 1862. See George E. Buckle, *The Letters of Queen Victoria* (Toronto: University of Toronto Press, 1926), 2nd series, vol. I, pp. 37–8.
36. Ibid., p. 38.
37. *Hansard*, 10 April 1862, cols 765–6. The skin referred to was a 5/8 inch of iron plate attached to the wooden backing of the target to represent the iron hull of the *Warrior*. That is, the total thickness of iron on the *Warrior* was close to 5 inches. However, it was customary to speak only of the thickness of the iron plating in front of the wood.
38. *Hansard*, 23 May 1862, cols 2148–54.

39. Ibid., col. 2154. A year later, when the First Lord of the Admiralty was asked about the results of a test of an Armstrong gun at Shoeburyness, Paget recalled his premature pronouncements, after which 'he had occasion very much to modify his expressed opinion'. This time he postponed comment until the target was properly examined. See ibid., 20 March 1863, col. 1653; also ibid., 23 June 1862, cols 843–4.
40. Ibid., 6 May 1862, col. 1281.
41. H.C. Matthew (ed.), *The Gladstone Diaries*, vol. 6 (Oxford: Clarendon Press, 1978), p. 126, entry for 3 June. Gladstone had made notes for a speech (ibid., fn. 7) but he did not participate in the debate. Cobden took delight in pointing out that the Chancellor of the Exchequer, whose opposition to the forts was well known, usually avoided the House of Commons when the issue was debated. (*Hansard*, 23 June 1862, col. 170). Gladstone insisted that his absence from the House was not related to the fortifications issue and he pleaded that, although he held a 'personal incompetency' to judge a military scheme, he stood 'pledged and committed as a Member of Her Majesty's Government' (*Hansard*, 10 July 1862, col. 176).
42. *Hansard*, 23 June 1862, col. 910.
43. Ibid., col. 947.
44. Quoted by Tennent, *Story of the Guns*, p. 168. This book, first published in 1864, provides a detailed history of Whitworth's work on rifling.
45. *The Times*, 27 Feb. 1860, p. 5; 28 May, p. 12.
46. *The Times*, 4 Nov. 1862, p. 12 (an article reprinted from the *Army and Navy Gazette*). See also Tennent, *Story of the Guns*, p. 298.
47. *The Times*, 6 Nov. 1862, p. 4.
48. Ibid., 10 Nov. 1862, p. 5.
49. Ibid., 13 Nov. 1862, p. 9.
50. See Anderson's testimony to the Select Committee on Military Organization, *Parl. Papers* VII (1860), qs 5716–81.
51. *London Review of Politics, Society, Literature, Art and Science* (14 Nov. 1862), pp. 423–4 and 6 Dec., pp. 490–91, 505. Whitworth's reply is included in the issue of 6 Dec. 1862, pp. 490–91. The *Review*'s final comment is on p. 505.
52. Rendel Papers, Tyne & Wyre Public Archives, Newcastle (hereafter, T&W), 31/4. This is the fourth letter in a collection of mostly in-letters from a number of men and is a valuable source for the history of Elswick. It begins in March 1863, when Stuart Rendel became involved with the firm, and ends with his death in 1913.
53. 'Report of the Select Committee on Ordnance', *Parl. Papers* XI (1863), qs 858–64.
54. T&W, 31/3372 (8 March 1863).
55. Ibid.
56. 'Report of Ordnance Select Committee', *Parl. Papers* XI (1863), qs 1131–54.
57. Ibid., qs 1204–9 and 1488–1509.
58. T&W, 31/3372.
59. 'Report of the Defence Commissioners on Spithead Forts', *Parl. Papers* XXVII (1862), qs 702–53, 309–23. For Whitworth's acceptance of Armstrong's challenge, see *The Times*, 19 Nov. 1862, p. 12.
60. *The Times*, 22 Nov. 1862, p. 12.

61. Rendel, *Personal Papers*, p. 273.
62. Ibid. The controversy to which Rendel refers involved an intense debate over the performance of the Armstrong breech-loading guns (not the muzzle-loading 300- and 600-pounders) which will be examined shortly.
63. Ibid.
64. Ibid.
65. Quoted by Tennent, *Story of the Guns*, p. 245.
66. Ibid., Part III, chapters 3 and 4. A good description of the Armstrong 600-pounder is in *The Times*, 21 July 1864, p. 12 and in Armstrong's letter a few days later, 25 July, p. 7, and his letter of 27 June, 1865, p. 14. See also F.S. Stoney and C.J. Jones, *Text-Book*, pp. 12–28. Ian Hogg, *The Illustrated History of Ammunition from the Beginnings to the Present Day* (Secaucus, N.J.: Chartwell Books Inc., 1985) is helpful.
67. T&W, 31/57; Armstrong–Whitworth Committee, 'Report to the Secretary of State' (6 Feb. 1864), Adm. 1/5905.
68. Wilmot to Armstrong, 24 Oct. 1864, T&W, 31/3681.
69. T&W, 31/60 (8 Nov. 1864).
70. T&W, 31/3389.
71. Ibid., 31/55.
72. About this time guns began to be named according to their calibre. Thus the Armstrong 600-pounder was also called the 13-inch gun.
73. *The Times*, 25 July 1864, p. 7.
74. Tennent, *Story of the Guns*, p. 308.
75. T&W, 31/19 (27 Dec. 1863).
76. *Illustrated London News*, 9 Jan. 1864, p. 37. This paper remained a vehicle for the company to display its new weapons. In 1876, it illustrated the Armstrong's 100-ton gun and, in 1877 and 1879, Elswick's newest gunboats. See *The Illustrated London News*, 10 March 1877, p. 233 and 16 Aug. 1879, pp. 159–60. The articles which accompanied the pictures were written by Stuart Rendel. See J.D. Campbell to Rendel, 7 Aug. 1879, T&W 31/4970 and 13 Aug. 1879, T&W, 31/3876.
77. *The Times*, 19 March 1864, p. 7.
78. Edgar Sheppard (ed.), *George Duke of Cambridge: A Memoir* (London, 1906), vol. 1, p. 249.
79. For example, *The Times*, 9 May 1865, p. 8; 27 June, p. 6; 8 June 1865, p. 11.
80. Rendel, *Personal Papers*, pp. 275–6.
81. *The Times*, 10 April 1862, p. 9.
82. Ibid.
83. I.S. Bloch, *The Future of War in its Technical, Economic, and Political Relations* (Boston: Ginn, 1899), p. 98.
84. *Hansard*, 11 April 1862, col. 855.
85. William L. Clowes, *The Royal Navy: A History from the Earliest Times to the Present* (London: Sampson Low, Marston and Company, 1903), vol. 7, p. 25.
86. Sir William White, who contributed much to the evolution of new warships, both at Elswick and at the Admiralty, describes in detail the series of transformations in marine engines which took place between 1860 and 1890 in his 'Address by the President', *Institute of Marine Engineers* (London: Harrison and Sons, 1893). A copy is in T&W 31/5287.
87. See Chapter 6.

88. D. Bonner-Smith and E.W.R. Lumby (eds), *The Second China War 1856–1860* (London: Naval Records Society, 1954).
89. Palmerston to Herbert, 5 Oct. 1859, in Stanmore, *Sidney Herbert*, II, pp. 298–9.
90. Grant to Herbert, 18 Aug. 1860, ibid., pp. 335–6; 8 Sept. p. 338.
91. Cambridge to Herbert, 3 Nov. 1860, ibid., p. 338.
92. *The Times*, 5 Nov. 1860, p. 6.
93. 'Report of the Select Committee on Ordnance', *Parl. Papers* (1863) XI, Appendix 48, p. 560. See also *Hansard*, 23 June 1862, col. 876.
94. The Armstrong gun used against the *Trusty* was an 80-pounder. For the *Warrior* test, Armstrong had increased it to the 110-pound size.
95. Armstrong sent many letters to the editor of *The Times* which outline the various trade-offs in gun construction. The above account is from his letter published on 10 April 1862. See also *Hansard*, 24 Feb. 1862, cols 681–2.
96. Stanmore, *Sidney Herbert*, II, pp. 406–10.
97. *Hansard*, 23 May 1862, cols 2151–4. See also the attack by Sir Frederick Smith and the response by Lewis, ibid., cols 2148–51.
98. Ibid., col. 2154.
99. *Hansard*, 6 March 1862, col. 1079.
100. Ibid., 4 April 1862, col. 626.
101. See the long letter from Armstrong to *The Times* on 25 Nov. 1861 which he published as a pamphlet, *The Armstrong Gun: Letter from Sir W.G. Armstrong to the 'Times', 25 Nov. 1861* (1861).
102. 'Report of the Select Committee on Ordnance', *Parl. Papers* XI (1863), p. v.
103. Clowes, *Royal Navy*, vol. 7 (1903), pp. 195–200.
104. *Hansard*, 10 March 1864, cols 1765–6. See also ibid., 18 April, cols 1179–82; *The Times*, 18 April 1864, p. 5 and 25 April, p. 14; Admiral Halsted, *England's Navy Unarmed* (London, n.p., 1864).
105. *Hansard*, 10 March 1864, col. 1779.
106. Ibid., col. 1781.
107. *Hansard*, 26 April 1864, cols 1624–5.
108. *The Times*, 15 March 1864, p. 12.
109. Armstrong to Rendel, 13 March 1864, T&W, 31/27 and 14 March 31/28.
110. Clowes, *Royal Navy*, vol. 7, p. 199.
111. 'A Copy of the Report of Admiral Kuper in reference to the Armstrong Guns in the Action of Simonosaki', *Parl. Papers* XXXII (1865), pp. 309–18.
112. *The Times*, 16 Jan. 1871, p. 4.
113. *Dictionary of National Biography* (1921), vol. 15, p. 118.
114. General John Adye, 'Memorandum by the Director of Artillery, on the Breech-Loading System of Field Artillery, April 1870', *Departmental Papers Relating to Muzzle-Loading versus Breech-Loading Rifled Field Guns*, W.O. 33/21A.
115. O.F.G. Hogg, *The Royal Arsenal* (London: Oxford University Press, 1963), vol. 2, p. 1417.
116. General John Adye, 'Memorandum'.
117. Ibid.
118. Paul Kennedy, *The Rise and Fall of British Naval Mastery* (London: Allen Lane, 1976), pp. 173–8; C.J. Bartlett, 'The Mid-Victorian Reappraisal

of Naval Policy', Kenneth Bourne and D.C. Watt (eds), *Studies in International History* (London: Longmans, 1967), pp. 189–208; R. Millman, *British Foreign Policy and the Coming of the Franco-Prussian War* (Oxford: Oxford University Press, 1965), pp. 148–58; Bernard Brodie, *Sea Power*, pp. 115–19, 160–67, 254–7, 440–45.

119. *Hansard*, 20 March 1863, col. 1709; see also cols 1757–70.
120. Stanmore, *Sidney Herbert*, II, p. 210. For Wilmot's replacement by Armstrong, see 'Minutes of Evidence, Royal Commission on Warlike Stores', *Parl. Papers* XV (1887), q. 2040 and Herbert to Armstrong, 4 November 1859, *Parl. Papers* XI (1863), p. 514.
121. Exact numbers are in Appendix 48 of the 'Report from the Select Committee on Ordnance', *Parl. Papers* XI (1863), p. 560. See also *Hansard*, 23 June 1862, col. 876.
122. For the correspondence, valuation lists and the final agreement, see Appendix 48 of the 'Report from the Select Committee on Ordnance', *Parl. Papers* XI (1863), pp. 567–99.
123. *Parl. Papers*, XXXII (1863), p. 221. For Rendel's role, see his *Memoirs*, p. 277 and Vicker's Archive, Document No. 70.
124. His resignation was generally welcomed. Lewis proclaimed that Armstrong's departure afforded an opportunity 'of correcting certain abuses of the old system' (*Hansard*, 20 March 1863), col. 1709. *The Times* interpreted Armstrong's resignation as unfortunate but the only way to resolve his conflict of interest. Armstrong immediately disputed that view in *The Times*, 13 Feb. 1863, p. 9 and 14 Feb. p. 12.
125. War Office orders for new guns from Elswick between 1863 and 1878 amounted to only £24 804.

PART II

MAKING THE GLOBAL ARMS MARKET, 1863–1914

Introduction

As Part I of this work has demonstrated, innovation in weaponry involves a complex entwining of technological, personal, professional and political factors. Engineers like Armstrong worked on armaments out of patriotism and the desire for professional reputation amongst their peers and the world beyond. Hopes for future commercial gains were no doubt present, but most of the entrepreneurs involved in the competition, and certainly Armstrong, were practising engineers with existing civilian engineering firms. Some of them no doubt saw technological innovation in armaments as another source of commercial gain; others, however, were attracted to armaments as a new and interesting technological challenge. This was certainly Armstrong's motive in 1854. Commercial considerations moved to the forefront later, and became the all-important driving force in armaments.

In other words, we should not too easily equate technological innovation with business entrepreneurship. A new technology, however brilliantly and elegantly designed, however efficiently it performs, does not necessarily guarantee being acknowledged as superior, nor does it necessarily bring business success. Armstrong demonstrated both technological and business entrepreneurship with his breakthrough invention of 1854 and its development over the following ten years. He successfully defended his invention as his own, and as the best design. His 1859 agreement with the British government was a brilliant move which ensured him the ability to develop his gun at Woolwich at government expense, and secondly he obtained a monopoly with a guaranteed stream of orders for the new private armaments company at Elswick. But his greatest moment in 1859 was the recommendation by the Royal Commission on the Defence of the United Kingdom that Britain defend its home shores from the land forts rather than the sea, a strategy which Lord Palmerston had appointed the commission to recommend on the basis of Armstrong's assurance that he would soon have the gun to place in those forts to justify the great expense of building those forts. But the most important event for the commercial success of his armaments innovations was no doubt the 1859 decision of the Royal Navy to put Armstrong breech-loaded guns on the new iron warship *Warrior*. Such an endorsement by the Royal Navy gave Armstrong an international reputation as the greatest gun builder of the time. When the 1859 contract was cancelled in 1863, it would be that reputation as the premier gun maker to the Royal Navy that propelled the Elswick Ordnance Company forward into a company of global dimensions.

Part II thus follows the attempts of Elswick to find and develop a world market for its guns, and its role in the proliferation, diffusion and transfer of armaments around the world through commerce. No one had any experience in this kind of commerce, and the company's early success came from Armstrong's reputation gained at home and the rising demand for armaments outside Britain. The name Armstrong was the company's most effective marketing weapon. Other guns were available, but the publicity gained by the Armstrong gun in England between 1854 and 1863 was beyond purchase, and set the Armstrong gun and Elswick apart from all competitors. The advantage of the Armstrong gun over that of Whitworth far outweighed the slight difference in design in the two guns.

Armstrong also moved into building small gunboats designed for his gun. All in all, the marketing entrepreneurship of Armstrong was built into his technology. As we shall see, the company relied on local agents to sell its armaments, rather than its own travelling salesmen: given the worldwide scope of the armaments market, and the absolute necessity for access to governments and war offices, local agents with such access were essential. It did sometimes put a distance between the company and the purchaser, but the benefits of commissioned agents far outweighed their disadvantages. Readers can judge for themselves in the following chapters whether or not Elswick's marketing activities and practices in the 1860s and 1870s were rational, efficient and effective.[1]

CHAPTER FOUR

World demand and Elswick's marketing

I

However complicated the invention and development of Armstrong's new guns, the demand for weapons in the third quarter of the nineteenth century was relatively simple: governments buy guns and warships to increase their ability to wage war.[2] The demand for guns came from war, and the second half of the nineteenth century had many wars. The Crimean War was the first in an age of war that would last for a hundred years, and that long era drove military technology to invent over and over a new ultimate weapon. We have seen the effects that the clash at Hampton Roads during the American Civil War and the War of Italian Independence had on the demand for guns, forts and ironclads in Britain. After 1860, many nations around the world experienced some kind of armed conflict. War freed peoples from their old masters, decided boundaries between new nations and revised boundary settlements of previous wars, while civil war decided the fundamental principles and values upon which new nations would develop.

During the 1860s and 1870s, war created a 'world' demand for industrial armaments: the American Civil War from 1861 to 1865; the Schleswig-Holstein War of 1864, the Austro-Prussian (Seven Weeks) War of 1866, which included naval battles between Italy and Austria; the continual civil war in Spain between 1868 and 1876 and her wars against Peru in 1864–5 and Chile in 1865–6; the war of Paraguay against Argentina, Uruguay and Brazil between 1864 and 1870; the War of the Pacific, 1879–84, which pitted Chile against Peru and Bolivia; and the continual wars between Turkey and the Balkan peoples which culminated in the Russo-Turkish War of 1877–8. China became a new customer for warships following the Japanese invasion of Formosa in 1874. World demand for weapons, especially for warships, increased during the 1880s and accelerated in the 1890s, when the great powers launched the armaments races and naval power entered a period of rapid growth and technological development.

The decade of the 1860s was, in short, a propitious time for Armstrong to enter the international armaments business. Armstrong guns were used in all the above conflicts. Armstrong embarked on this venture on the first day of January 1864, when the Elswick Ordnance Company was wound up and a new business, the Sir William G. Armstrong & Company, was formed to manufacture Armstrong guns for the markets outside Britain. Elswick entered the world arena early and played a leading part in the international armaments industry which emerged during the second half of the nineteenth century. Armstrong built larger and more powerful rifled muzzle-loaders during the 1860s and 1870s and, in the late 1870s, new propellants and improved steel enabled the company to manufacture large breech-loaders, the guns which laid the foundation for the great battleships of the late-Victorian and Edwardian era. The company began building gunboats in 1867, cruisers in the early 1880s, and constructed its first battleship in 1885 for the British Admiralty. Elswick sold Gatling machine guns under licence after 1870 and developed an enormously successful quick-firing naval gun in the 1880s. Its gun mountings were famous.

Competition was less intense in the 1860s and 1870s than after 1880. Whitworth sold guns abroad but he was not a serious threat. French manufacturers were prohibited from exporting heavy guns until 1885.[3] The British government, on the other hand, never questioned Armstrong's right to sell his guns to foreign governments who were not at war.[4] The main competitor on the international scene was Krupp, whose artillery was in great demand after the Franco-Prussian War. However, Armstrong remained the dominant manufacturer of quality naval armaments. Some non-industrial nations established their own arsenals and dockyards, but lack of an industrial base, capital and technological knowledge frustrated these attempts at independence.[5]

Throughout 1864, orders came in from Denmark, Sweden, Norway, Portugal, Spain, Italy, Chile, Brazil, Argentina and Peru. Foreign merchants and traders whose vessels sailed pirate waters purchased guns from Elswick.[6] In September, only nine months after the company had been formed, Armstrong was pleased to inform Rendel that 'the Danes have ordered 2 more 150 pders' and he was confident that Turkey could be set down 'as completely booked for guns and ammunition to the extent of 60 or 70,000£'. Elswick was now 'so well off for orders' that Armstrong felt they 'could be cool about Russia for awhile', although he suggested Rendel keep his plans to go there.[7] The Turkish order was soon increased to £150 000 and the gun factory approached full capacity.[8] On the other hand, the Spanish government was liable to fall at any moment (and did so in 1868) and Armstrong was concerned that a new government might not honour payments owing to the company. Fortunately, Armstrong did not have to accept the risk. As he told Rendel:

> The Works are getting so full of orders that the next difficulty will be how to deal with those that we can't execute. Under these circumstances I don't think the present is the time for *forcing* markets, and Spain, considering the present political aspect of affairs is about the last one to be so treated. I am therefore very doubtful of your taking the initiative in any move for new orders for Spain.[9]

The existing orders of £65 000 were not affected.[10]

Turkey became an early customer ('so far as she can pay') along with Egypt. The British colonies, especially Australia, also developed into 'not inconsiderable' markets. Denmark's orders between 1864 and 1867 were 'considerable' in view of that nation's small size, and Elswick was 'firmly established' in Norway. Holland was a small but new customer.[11] Even France bought a few 'experimental guns' to keep abreast of any new technology.[12] The Russians were also extremely interested in the Armstrong built-up gun. When Krupp's guns performed badly, the Russian government ordered a 300-pounder from Elswick and for a moment it was possible that Elswick could move into the Russian market. The Russians proposed that Armstrong erect a plant in Russia to replace both Krupp and their own factory at Alexandrossky but, as Rendel explained, Sir William was a cautious man and, with his 'heavy undertakings at home', he did not want 'to advance capital and incur responsibilities' in Russia. He was willing, however, to supply a Russian factory with machinery and personnel.[13]

Austria ordered guns for five ironclads it had built after the battle of Lissa in 1866. Her opponent in that naval battle, Italy, bought £100 000 worth of guns during 1865 and 1866.[14] Italy became the most important customer for Elswick's large naval guns and purchased the enormous 100-ton muzzle-loaders developed by Armstrong in the 1870s.

Elswick thus rapidly developed a firm position in the rapidly expanding global market for armaments and Armstrong's initial reluctance to sell abroad gave way to enthusiasm. A particularly jubilant day was 19 March 1864. Rendel's Egyptian orders were confirmed in the morning mail and another followed 'to the tune of near £4000 and proposals for more'. Finally, *The Times* contained a 'very satisfactory' article which described the unqualified success of the 600-pounder and its new steel shell against John Brown's 11-inch plate at Shoeburyness.[15] Most importantly to Armstrong, the trial demonstrated the efficacy of steel shells which were only a little more than half the weight of iron projectiles and required far less charge. The steel shell had not torn the rifling of the gun, one of the major questions the test was designed to answer. The only questions remaining were the endurance of the gun and whether or not a ship could be designed to carry it. This news came only a week after Lord Clarence Paget had criticised the 110-pounder for its performance at Kagoshima and prompted Armstrong's reply that more attention should be given to his successful muzzle-loaders. The report on the Shoeburyness test, said Armstrong, was 'a capital sequel to the Paget correspondence'.[16] Armstrong contemplated the place of Rendel in the future of the firm:

> We must [he wrote] make your services better worth your time this year and I think you will do wisely to identify yourself with the Elswick Company which will most probably increase in importance and which if your usefulness continue will probably in due time find it advisable to form a permanent connection with you.[17]

Rendel purchased a twenty-fifth share in Elswick for £13 000 in 1870,[18] a position that grew over the years until he became a major shareholder.[19]

Demand for armaments was limited by the financial strength of governments and their commitment to other budgetary items. The cost of guns and warships increased over the years, as did the costs of maintaining trained gunners and military and naval organisations required by modern industrial armaments. In the 1850s, the total cost of the 130 guns and carriages for a first-class warship was £20 000. The projectiles cost about £1 each. In the 1880s, the 12 large guns on board a battleship cost £200 000 and projectiles cost up to £240 each.[20] In the 1880s, torpedoes cost £500 each. In 1877, one Armstrong 7-inch, 6¾-ton rifled muzzle-loader cost £925 and its carriage £570. The small 40-pounder cost £330 and its carriage £275. The smallest Elswick *Staunch* class gunboat, with guns, cost £25 000 in 1877.[21] After 1890, the cost of one battleship approached £1 million; after 1905, Dreadnoughts cost nearly £2 million each and about £500 000 per year to maintain. One shell could cost more than £1000.

War was expensive. The brief campaign to China in 1860–61 cost Britain £3 800 000.[22] Even during the world economic expansion of the 1860s and 1870s, small nations spent a large portion of their revenue on armaments. The depression of the 1880s and 1890s increased the budgetary pressure and South American republics, for example, spent as much as half their revenues on armies and navies. In 1890–91, the Argentine government defaulted on its debt as a consequence of the revolt in Buenos Aires. Brazil, Argentina and Chile directed 16 per cent, 19 per cent and 25 per cent of their budgets, respectively, to military and naval expenses in 1905.[23] In 1908, Argentina spent more than $20 million for two American battleships in its naval arms race with Brazil,[24] while Brazil negotiated with Elswick for similarly-priced British battleships.[25] In the early 1890s, China's central government spent 22 per cent of its annual revenue on its army and navy.[26] Italian finances were also severely strained in the 1890s by her large purchases of armaments.[27] Armstrong's financial difficulties with Turkey are examined in the next chapter.

War affected the demand for guns and warships in another way. War was the ultimate proving ground for weapons, and success in battle was the most effective advertisement. The demand for Krupp artillery (as well as the Prussian military organisation) increased dramatically after the Franco-Prussian War, while the performance of Elswick weapons (especially its quick-firing guns) in the Sino-Japanese War of 1894–5 reaffirmed that company's reputation as an innovative manufacturer of sophisticated high-quality guns.[28]

Elswick's reputation was also enhanced by its association with the British navy. The Admiralty was not allowed to buy guns from Elswick after 1863, but the Royal Navy continued to carry guns made by the Armstrong built-up method, for which the Woolwich Arsenal held the patent rights donated to it by Armstrong in 1859. Although Armstrong managed to sell only a few *Staunch* gunboats to the British Admiralty in 1867, the sale represented the stamp of approval of the world's foremost navy. Elswick would refer to its work for the British Admiralty in its tenders to foreign governments.[29] The company also benefited from being part of Britain's great arsenal. In 1881, when a Chinese admiral set off for England and Europe, Elswick's agent in China reminded the company – unnecessarily no doubt – that 'the more the Admiral is impressed with the greatness of England's Dockyards, Arsenals, Factories etc. the less likely he is to fall into the views of Li Feng Pao, who is sure to endeavour to bring him over to the Krupp interest'.[30] Li Feng Pao was China's minister in Germany until he was dismissed in 1884 on suspicion of receiving a substantial personal commission (that is, bribe) for ordering German ships for Li Hung-chang.[31] When he arrived, the admiral was given the grand tour: presentation at Court, visitations to the royal dockyards and Woolwich with the usual banquets and complimentary speeches.[32] The admiral was no doubt impressed by his visit to the Elswick factories and dockyards, which had spread out along the River Tyne. The firm reminded customers, as they did the Peruvians in 1885, for example, that Elswick had 'exceptional experience in warship construction and armament' and emphasised that their 'capability of dealing with the whole of the works including the armaments would ensure both economy and rapidity in the completion of the vessels'.[33]

II

The British royal treatment and displays of Elswick's technological grandeur may have impressed visitors, but German royalty and Krupp factories were also intoxicating, and armament sales depended upon local agents in the various markets. When Armstrong returned to Newcastle in 1864 to manage the new company and continue his research and development, Stuart Rendel remained in London to represent the firm on the Armstrong–Whitworth Committee and as Elswick's sales manager and negotiator, for which he received 5 per cent commission.[34] Foreign governments followed the Armstrong–Whitworth competition closely in the newspapers, but came to Rendel for additional information. The Russian Ambassador in Paris, who followed the trials 'day by day', wrote to Rendel that he could not give 'a full confidence to these articles as I know that they are not *official*' but 'merely all what the correspondents have been able to collect'. He thus requested that Rendel send him 'a copy of *a very true* and if possible *official* report on the present experiments'.[35] 'No person better than you,' he later added, 'can give me the wanted information.'[36]

The above discussion opens another area of debate about the performance of British Victorian entrepreneurs: the accusation that they failed to sell their wares in foreign markets aggressively compared to their German and American competitors. British businessmen, the argument goes, refused to adapt their domestic products to the particular needs or desires of their foreign customers, did not employ salesmen who spoke the local language, and scorned the small order.[37] British technological innovators, that is, too often failed in selling effectively their inventions in the global market. They too often believed that the best technology would capture its demand and that there was one technology that was superior to all others. British entrepreneurs are accused of falsely assuming that a particular engineering problem had a single best solution which was universally applicable. Accordingly, a technological innovation that proved the best solution in one part of the world would also be the best technology everywhere, a belief that led to the 'take it or leave it' attitude. Demand, being rational, would take it since it was the best product. Thus British manufacturers took little interest in the techniques of salesmanship. It was this belief about the nature of technology, and its relationship to society, that led them to view marketing as an unnecessary expense to be turned over to commissioned agents or merchant houses which served many clients and had less incentive than a salesman who worked directly for the company.[38] British entrepreneurs thus built an organisational and a communication barrier between themselves and their overseas customers.[39]

The above failures may or may not indicate anti-entrepreneurial acts. It may well be that local agents and established merchant houses were the best way to market armaments. It is certain that Armstrong's technology did create its own demand because it was judged the best by authorities of high esteem in the most respected navy in the world, and customers sought out Armstrong because his guns, gun mounts and innovative warships were leading technologies and proved effective in all parts of the world. At the same time, Armstrong knew as well as anyone that there was more than technology to selling armaments. To repeat an important point, this book was not written as another test case for the thesis of the failed British entrepreneur. One can make two lists, with all of Elswick's entrepreneurial transgressions on one side and all the brilliant moves and successes on the other, and then make a judgment. The problem for Armstrong is that the list has two columns and the case cannot be decided without ignoring too many items. But beyond that is the question of the historian's objective, and the purpose here is to tell the Armstrong story, to set Armstrong's policies and actions in their context in order to understand more fully the undoubtedly central role played by Armstrong and his Elswick company in the rise of the modern armaments industry and the global proliferation of armaments in the nineteenth and early twentieth centuries.

In the beginning, Rendel generated a number of orders from London, but as the company's business spread throughout the world, resident agents abroad became essential. The proven quality and power of the Armstrong built-up gun, the publicity surrounding the trials supervised by the British War Office in the early 1860s, and the growing demand for guns created by the increasing number of military conflicts throughout the world, attracted many individuals anxious to represent the firm abroad.[40] Overseas agents or 'ambassadors' with access to foreign political and military establishments were essential to the successful marketing of weapons. The man on the spot required a thorough knowledge of the local political environment, access to the powerful few who procured armaments and the ability to move quickly whenever an opportunity to make a sale arose. Honesty, loyalty and some knowledge of weapons technology and its relationship to naval strategy and tactics were also important characteristics of an effective agent.

These requirements posed different problems for marketing weapons in Europe and in the rest of the world. The company directors and Stuart Rendel developed direct contact with European governments and military establishments since these were close at hand and access to the military–political hierarchies was readily available. Britain and Europe were ancient neighbours with common historical roots; London was full of European diplomats and foreign military officers (many of whom were invited to witness the Armstrong–Whitworth trials) and the intricacies of European political developments and diplomatic alignments which affected armaments sales were quickly known and understood. There was no similar intimate understanding between Britain and China, for example. No businessmen or military attaché in London could provide much assistance to the company in clearing a path to the highest levels of the Chinese power structure, and a permanent Chinese legation did not arrive in London until 1877. On the other hand, while European markets were the largest in the world, competition from domestic producers was virtually insurmountable: selling Armstrong guns to Germany, Austria or Russia was not the same as competing with Krupp in China.

Some of Elswick's early business depended upon the financial entrepreneurship of its agents. In 1881, an Armstrong agent, John Pitman, was certain of an order for two 25-ton and five 15-ton guns for Canton but was forced to finance it himself with the help of Butterfield and Swire, a large trading house in Asia.[41] Thomas Glover & Company, Elswick's first agent in Japan, obtained substantial orders in the 1860s by lending the Satsuma the money to buy guns for the struggle against the Bakufu, the central Tokugawa military government. Glover was also able to sell Elswick guns to the Bakufu despite its political attachment to France.[42] The Newcastle partners did not consider these financial transactions as any of their business.[43] The company firmly maintained its policy of one-third down, one-third during production and one-third upon delivery. Glover finally overextended himself and went bankrupt.

Agents without thorough technical knowledge of weapons created difficulties, although the company must share some of the responsibility because of its failure to supply its agents with adequate technological advice and not investigating obvious errors in orders. In 1865, Elswick sent guns to Japan without loading mechanisms or shells simply because Glover had not included them in the order. As Glover wrote to Elswick, 'owing to our want of experience in these matters we were quite ignorant when sending forward the orders, of the vast quantity of stores etc. necessary to be furnished with the guns & ammunition'.[44] Twenty years later the company still had difficulty providing its Asian agents with full technical support. Alexander Michie of Jardine, Matheson complained in 1886 that Elswick wanted to know 'exactly what kind of torpedo boat the Chinese require'; but, he said, 'that is just what the Chinese do not know themselves'. The absence of technical advice from the company exacerbated the slowness in sending technical documents to Jardine.[45] One agent, Bridgford, was not kept fully and immediately informed of Elswick's prices.[46] Robert Thompson ended his relationship with the firm with a lawsuit for non-payment of commissions and expenses.[47] Asian agents often knew much of Asia but nothing of armaments, while Elswick's partners knew everything about armaments but little of Asia.

The firm principally relied upon J.J. Bucheister of Shanghai and Bidwell, an independent agent who worked through either Bucheister or Jardine, Matheson & Company, which became Elswick's chief representative in Asia in the 1880s, but accepted orders from anyone.[48] The purchase of gunboats for China negotiated through Sir Robert Hart, the Inspector General of the Chinese Maritime Customs, drew protests from Sharpe & Company, the 'official' agent of the company in Hong Kong. Hart himself agreed that it was unwise of Elswick to circumvent their own agent who presumably sought orders continually and actively while he, Hart, did not.[49] 'I suppose [he wrote to his London representative] the manufacturer has said to the agent: "pretty fellow you are not to catch it when there *is* an order given!"'[50] In 1886, Jardine, Matheson demanded that Elswick refuse a Chinese order from an American firm, Russell & Company,[51] and sparked a debate within the firm over its policy towards its agents. One director argued that Matheson had 'utterly failed to represent us themselves in this very important business' and insisted that, since Jardine was not Elswick's sole agent, Russell ought to be given the order.[52] After much debate Jardine was confirmed as Elswick's main Asian agent, but Bucheister was also allowed to continue to represent the firm.[53] Thus Elswick relied upon a system of commission agents rather than developing a well-defined organization of employee-agents.

Reports to the company in the early 1880s describe very busy agents in Asia but few sales. The drop in new orders cannot be blamed entirely on poor marketing. Sales tactics were progressive: photos of Armstrong guns with descriptions in Chinese were sent to all the port authorities and letters to friends requesting help. But when demand evaporated there was little one could do. One agent reported that, although his letters of introduction from Elswick gained him access to important people in Asia, after a year the prospect for orders remained poor.[54]

An agent had to balance his role as the representative of a foreign company with his career inside the country to which the arms were sold. Robert Hart was busy with many personal investment schemes as well as his heavy responsibility for the Chinese Customs Service.[55] However, his commitment to Elswick and the promotion of British interests in China (which he equated with China's best interests) was sincere and, as we shall see, his enormous political influence contributed to the company's position in China. Others too worked on behalf of Armstrong but linked their futures with China, not the company. Sir Halliday Macartney supported Armstrong, but his career hinged on Tseng Kuo-fan and Li Hung Chang. When Tseng died and Li was forced to sacrifice Macartney over an explosion of two guns at the Taku Forts, Macartney's career in

China was over and Armstrong lost an advocate. Macartney returned to England as the secretary of the first Chinese Legation.[56]

In the early 1880s Jardine, Matheson & Company became an Elswick agent in China. Jardine, Matheson was an old and successful British company which had been promoting trade and investment in China since the early nineteenth century, having established itself in the Chinese tea and opium trade after the East India Company lost its monopoly in 1834. The company had avoided any direct involvement in selling arms – perhaps because they feared it might result in bad relations, as had happened with the opium trade, perhaps because it went against the conscience and beliefs of senior members of the firm like Hugh M. Matheson, who, in 1843, had refused to represent the firm in China because of its involvement in the opium trade – but had extended credit to Glover and other arms agents and provided loans to China and Japan for military purposes in order to strengthen the political position of the firm.[57] Jardine, Matheson strengthened its political position by lending China money for its northern campaigns against the Muslims. Jardine viewed arsenal and warship projects as one of the first steps towards the industrialisation of China in which it would claim the place as the predominant middleman between China and western industrial interests.[58] Thus, whether it was a consequence or a cause of Glover's bankruptcy, or a result of Hugh Matheson's devoting all his attention to a Spanish mine company, after 1880 Jardine, Matheson became directly active in the arms business as an agent for Elswick.

Bribery was common in China, and arms sales could generate 'commissions' as high as £50 000.[59] In Japan, where a similar system operated, a Krupp sub-agent and vice consul acquired 'commissions' of $250 000 in a short time.[60] Hart would have none of this. He not only refused to employ bribery as a means of obtaining orders but he also refused to accept the commissions offered by Elswick.[61] He preferred to use them to lower the price paid by the Chinese government for the gunboats. China's Mandarin class depended on bribes, and Hart's refusal to accommodate them was a threat. A Mandarin faction 'plotted against him ... [and tried] to damage the Viceroy's faith in ... these new acquisitions ... and Mr. Hart's advice'. Hart was able to withstand this pressure and convinced Li Hung-chang, the Viceroy of Chihli and the man in charge of China's armaments purchases, to buy Armstrong cruisers instead of Krupp ironclads.[62]

Armament sales depended upon the ever-shifting political ground. Inside all governments there were political parties and military and naval factions who demanded Armstrong guns and others who demanded Krupp arms; others insisted a domestic industry be established; and small 'peace parties' wanted no new armaments at all. Armaments agents and ambassadors paid close attention to the ebb and flow of these political struggles, ready for immediate action should any political change create an opportunity for a sale. As Bridgford pointed out in his 1881 report, agent Edward Whittall was 'working away at the Japanese in favour of our guns', but more important was the possible forced retirement of a prominent Krupp advocate and the coming to power of the opposition party which favoured Armstrong. If this occurred, said Bridgford, Whittall would immediately apply to have Armstrong guns tried against those of Krupp.[63]

Agents with good political and military contacts were hard to find. One source was the pool of foreign correspondents for *The Times*. Robert Thompson, for example, worked for both *The Times* and Elswick between 1886 and 1894, and for Armstrong alone until 1897. Thompson had wide knowledge of politics in Spain, Portugal, South America and Asia, and was personally acquainted with highly-placed officials in those states. But his relationship with the firm was at times vague and shifting. At one time he was an official, commissioned agent with a concrete contract,[64] at another time his status with the company 'was somewhat analogous ... to that of a private diplomat or ambassador' whose role was 'to find out what was happening in various foreign countries, to let his employers know what was likely to be required, and generally to prepare the ground for orders for warships and war material'.[65] This ambiguity led to the lawsuit over whether or not certain activities constituted commission work. The work on behalf of Elswick by the British army officer and *Times* correspondent, Charles Brackenbury, is related in full in the next chapter.

Agents could earn enormous commissions and there were always plenty of applicants. Unsolicited offers to represent the firm arrived continually from individuals around the world who explained how their unique position would benefit the company if only they were recognised as Elswick's agent. Others came from Armstrong ambassadors who wanted to convert their status into that of a commissioned agent. The Brazilian ambassador presented his case in 1882. He claimed that he had obtained 'the useful intervention of our friend Admiral Hilvina[?] da Motta', after which the Brazilian Admiralty had unanimously agreed to cancel a contract with Whitworth for naval guns and throw all their business to Armstrong. The ambassador carefully described how his name was well known in this matter and that the Brazilian Admiralty considered it necessary that he become Elswick's official agent.[66]

Thus, while war, arms races and international power politics generated a global demand for weapons in the 1860s and 1870s, the share of that demand which came to Elswick depended upon well-placed agents and ambassadors. Many agents saw Elswick as a means to accomplish goals and fulfil ambitions unrelated to the firm's interests. Their purposes and characters differed widely and they affected the company's performance in the various markets of the world in different ways. Elswick's experience in the United States, Europe and Asia provides revealing illustrations of the way the company responded to the growing demand for weapons in three different areas of the world.

Notes

1. The marketing practices of British entrepreneurs have been much criticised. For a brief summary of that aspect of the debate, see P.L. Payne, *British Entrepreneurship in the Nineteenth Century* (London: Macmillan, 1974), pp. 41–4, 53–5.
2. In the 1890s, other factors such as imperialism, national prestige, domestic economic and political pressures and international rivalries became significant factors in generating the demand for armaments.
3. William H. McNeill, *The Pursuit of Power: Technology, Armed Force, and Society Since A.D. 1000* (Chicago: University of Chicago Press, 1982), p. 240.
4. *Hansard*, 2 March 1865, cols 960–61.
5. Japan is the great exception, but even its military and naval factories were built with German and British capital and expertise. A large portion of the Japanese navy that destroyed the Chinese fleet in 1894 and the Russian convoy in 1904 was built at Elswick.
6. *Parl. Papers* XI (1863), p. 510.
7. Armstrong to Rendel, 7 Sept. 1864, T&W 31/50. See also 31/59 for similar references to full capacity.
8. Armstrong to Rendel, T&W, 31/56 and 31/59.
9. T&W, 31/224 (7 April 1867).
10. T&W, 31/241 and Rendel to Hay, 3 June 1867, T&W, 31/1934.
11. Rendel makes a curious reference about much petty trade of the firm with 'outlandish powers', adding that 'as these Governments are generally in the hands of merchants and commission agents who are indifferent to anything but profit we have very little to do with them' (T&W, 31/1934, pp. 19–20).
12. The French were perhaps more interested in the hydraulic machinery with which Armstrong guns were manipulated, an aspect of large guns which became increasingly important. See the 'Royal Commission on Warlike Stores', q. 7518–24. Evidence of Rendel's key role in what success Elswick had in France is in T&W, 31/23 and 31/25.
13. Armstrong to Rendel, 7 Feb. 1866, T&W, 31/138. Rendel's letter to Hay was confidential, but Sir John, perhaps with Rendel's approval, told Parliament about the Russian offer (*Hansard*, 13 June 1867, col. 1797).
14. J.D. Scott, *Vickers: A History* (London: Weidenfeld & Nicolson, 1962) p. 23.
15. *The Times*, 19 March 1864, p. 7.
16. Armstrong to Rendel, 20 March 1864, T&W, 31/33. The 'Paget correspondence' is in *The Times*, 15 March 1864, p. 12. For further debate over the Armstrong 100-pounder at Kagoshima, see *The Times*, 30 Oct.

1863, p. 6; 25 Nov. 1863, p. 8; 18 April 1864, p. 5; 25 April 1864; Oscar Parkes, *British Battleships, "Warrior" 1860 to "Vanguard" 1950: A History of Design, Construction and Armament* (London: Seely Service, 1966), p. 34; *Hansard*, 25 Feb. 1864, col. 1123; 10 March 1864, cols 1778–80 (Paget's attack), 18 April 1864, cols 1179–82, 26 April 1864, cols 1624–5. The discussions among Armstrong, Noble and George Rendel are described by Armstrong in T&W, 31/27, 31/28, 31/43 and 31/44.

17. T&W, 31/33.
18. T&W, 31/1942 (3 Feb. 1870).
19. In 1893, he held 103 881 £1 ordinary shares and 11 860 preferred shares; in 1898, he owned 256 900 shares which generated over £30 000 income. See T&W, 31/3677, /4031, /6592, /6597 and /6602. In 1911, Rendel's common shares were 191 000 (31/4724).
20. 'General Memorandum dealing with the Growth of Expenditure on Vote 12 of the Army Estimates, 1870–71 to 1886–87', Appendix, 5th 'Report of the Select Committee on Army and Navy Estimates, 1887', *Parl. Papers* VIII (1887), p. 812.
21. 'Estimate by Sir W.G. Armstrong & Co. for one 7-inch and six 40-pounder M.L. guns with Carriages', 28 May 1877, T&W, 31/4608.
22. *Parl. Papers* XLI (1860), p. 313.
23. Percy F. Martin, *Through Five Republics (of South America): A Critical Description of Argentina, Brazil, Chile, Uruguay and Venezuela in 1905* (London: William Heinemann, 1905), pp. ix–x, 353–5.
24. Thomas F. McGann, *Argentina, the United States, and the Inter-American System 1880–1914* (Cambridge, Mass.: Harvard University Press, 1957), pp. 267–9.
25. D'Eyncourt Papers, Brazil File. See also *The Economist*, 5 Oct. 1912, p. 625.
26. Calculated from Albert Feuerwerker, 'Economic Trends in the Late Ch'ing Empire, 1870–1911', in John K. Fairbank and Kwang-ching (eds), *The Cambridge History of China*, vol. II, *Late Ch'ing, 1800–1911*, Pt 2 (Cambridge: Cambridge University Press, 1980), Table 21, p. 63.
27. C.J. Lowe, *Salisbury and the Mediterranean, 1886–1896* (London: Routledge & Kegan Paul, 1965), p. 78.
28. *Arms and Explosives*, Oct. 1895, p. 6.
29. Elswick to Peruvian Consul in London, 31 Jan. 1885, T&W, 31/2689. In 1900, the chairman of the Thames Ironworks and Shipbuilding Company told the shareholders that 'the advantages of Admiralty work were that it put a company in the front rank and enabled them to secure foreign contracts' (quotation from Arthur Marder, *The Anatomy of British Sea Power: A History of British Naval Policy in the Pre-Dreadnought Era* (New York: Alfred Knopf, 1940), p. 26, fn. 3).
30. Bridgford report (23 Dec. 1880), T&W, 31/4192.
31. Kwang-ching Liu, 'The Military Challenge: the North-West and the Coast', in John K. Fairbank and Kwang-ching (eds), *The Cambridge History of China*, vol. II, *Late Ch'ing, 1800–1911*, Pt 2 (Cambridge: Cambridge University Press, 1980), p. 255.
32. Demetrius C. Boulger, *The Life of Sir Halliday Macartney* (London, 1908), p. 324.
33. Letter to Peruvian Consul, 31 Jan. 1885, T&W, 31/2689.
34. T&W, 31/3433 ([9?] April 1864).

35. T&W, 31/3431 (8 April 1864).
36. T&W, 31/3438 (29 April 1864).
37. Ernest Williams, *Made in Germany*, 2nd edn (London: Heinemann, 1896).
38. P.L. Payne, *British Entrepreneurship*, pp. 41–5.
39. Other charges against late nineteenth-century British entrepreneurs include clinging to the ideals of individualism which led to their refusal to amalgamate, integrate or establish strategic alliances with each other to compete against the large German cartels and the huge American corporations.
40. T&W, 31/3437, /2787, /6155, /3260 and /3263.
41. T&W, 31/5277.
42. Meron Medzini, *French Policy in Japan During the Closing Years of the Tokugawa Regime* (Cambridge, Mass.: Harvard University Press, 1971).
43. Shinya Sugiyama, 'Thomas B. Glover: A British Merchant in Japan, 1861–70', *Business History* (July, 1984), pp. 115–38.
44. Ibid., p. 123.
45. Edward Le Fevour, *Western Enterprise in Late Ch'ing China: A Selective Survey of Jardine, Matheson and Company's Operations 1842–1895* (Cambridge, Mass.: Harvard University Press, 1968), pp. 87, 184 (fn. 29).
46. T&W, 31/2967.
47. The company settled out of court. See *The Times*, 16 Dec. 1904, p. 3; Francis W. Hirst, *The Political Economy of War* (London: Dent, 1916), pp. 98–102.
48. Le Fevour, *Western Enterprise*, p. 184, fn. 29; John K. Fairbank, Katherine Bruner and Elizabeth Matheson (eds), *The I.G. in Peking: Letters of Robert Hart, Chinese Maritime Customs 1868–1907* (Cambridge, Mass.: Harvard University Press, 1975), vol. 1, letter 102, fn. 1. See also T&W, 31/2725.
49. John K. Fairbank et al., *The I.G.* vol.1, p. 207 (letter 139).
50. Ibid., letter 131.
51. T&W, 31/2715 (7 April 1886).
52. T&W, 31/3120 (9 April 1886).
53. Dyer to Rendel, T&W, 31/2725 (11 April 1886).
54. T&W, 31/2974. The Bridgford reports are T&W, 31/4667 (8 Dec. 1880), 31/4192 (23 Dec. 1880), 31/5277 (25 Oct. 1881), 31/4187 (1 Nov. 1881), and 31/4190 (17 Jan. 1882).
55. T&W, 31/4190.
56. Boulger, *Life of Macartney*, pp. 78, 124–5, 232–41, 278–9, 432; Stanley F. Wright, *Hart, and the Chinese Customs* (Belfast: William Mullan & Son, 1950), p. 477.
57. Le Fevour, *Western Enterprise*, p. 62.
58. Ibid., pp. 62, 89; see also Charles Harvey's biography of Hugh M. Matheson in the *Dictionary of Business Biography* (*D.B.B.*) (vol. 4, pp. 187–91).
59. Campbell to Hart, Campbell Papers, Ms. 191931, Box 7, Envelope 30, Z Letters, No. 2 (Jan. 15, 1875).
60. T&W, 31/1935.
61. Campbell wanted to accept them, but Hart said no. Campbell to Hart, Campbell Papers, Ms. 191931, Box 7, Envelope 29 (26 Oct. 1877). The 'special line' offer from Noble is in T&W, 31/4892.
62. Bridgford to Elswick, T&W, 31/4667 (8 Dec. 1880).

63. T&W, 31/5277.
64. Two and a half per cent on hulls and engines, 5 per cent on other war materials for Chile; 1 per cent on all armaments ordered by China and Japan through him; see *The Times*, 16 Dec. 1904, p. 3.
65. Ibid.; see also F.W. Hirst, *Political Economy*, pp. 98–9.
66. Victor Hias[?] to W.G. Armstrong & Co., T&W, 31/5875 (May 1, 1882).

CHAPTER FIVE

Arming America and Europe in the age of nationalism

While the Crimean War of 1854–6 prompted Armstrong to design his new artillery piece, and the French rearmament of 1858–63 opened the British treasury for him to develop his small weapon into a monster gun, nationalist wars in America and Europe ensured the commercial success of Elswick. The nationalist wars of the 1860s in America and Europe accelerated the industrialisation and mechanisation of war and the commercialisation of armaments production for a global market. Armstrong skilfully rode these global industrial waves which lifted his company from the depths of bankruptcy to the heights of financial success.[1]

I

Amongst Elswick's first customers were each side in the American Civil War. In the north, a vigorous and innovative small-arms industry was an important part of the early industrialisation process in the United States during the first half of the nineteenth century[2] and, when they turned their attention to artillery, Americans like Dahlgren and Rodman were soon recognised as leaders amongst the world's innovative artillery engineers. But the Civil War created a demand for guns far beyond American productive capacity, and the American market opened up to British gun manufacturers like Armstrong, Whitworth and Blakely.

In 1861, the Armstrong gun was well known as the choice of the British army and navy, but selling weapons to either side in the war was a delicate matter for British arms manufacturers, given the Cabinet Order-in-Council of 1861 which prohibited the export of arms to the belligerents.[3] It did not, however, prevent Armstrong, Whitworth or Blakely from selling guns to each side.[4] When the agent for the Massachusetts government came to London to purchase Armstrong rifled muzzle-loaders to protect Boston harbour, he wrote back to Governor Andrew that 'the business must be managed with great secrecy and caution' since the British government would never allow such a transaction and, if the sale became known, a 'great public outcry' would follow, particularly from Armstrong's political enemies inside Parliament and those in the War Office who supported Whitworth.[5] The agent, Colonel Harrison Ritchie, thus negotiated with Elswick through a third party, John Scott Russell.

Russell was an eminent naval architect in whose shipyard Brunel had built his *Great Eastern*. He had worked on the *Warrior*, was a founding member of the Institute of Naval Architects and an active participant in the debates over the kind of warships Britain should build. In 1863, he attended the annual meeting of the British Association in Newcastle, at which Sir William Armstrong delivered his presidential address and conducted a tour of the Elswick plant. Russell met Andrew Noble, the artillery officer whom Armstrong had recruited from the British army and made a full partner at Elswick. Noble informed Russell that Elswick was seeking foreign customers and presented Russell with the opportunity to earn a commission as middle man between the company and an American government. In January 1864, he called upon the American minister in London, C.F. Adams, who sent him to Colonel Ritchie. With this fortuitous timing, Russell became the agent of the Massachusetts government for the purchase of guns from the Sir William G. Armstrong and Company. He went back to Noble who agreed to pay him a commission of 5 per cent of the selling price of any orders he brought to the company.[6] Russell received a £5000 first instalment from Adams, out of which he was to pay Elswick for guns as invoices for completed work were sent to him by the company.[7] Elswick would thus claim it sold guns to Russell, not a belligerent in the American Civil War.

This was a tremendous opportunity for Russell to establish a profitable relationship with Elswick. His ship building business was on the verge of bankruptcy but a new career with a promising armaments company lay before him. However, his vanity, ambition, poor judgment and debts ruined this opportunity and his reputation. He made payments up to £8734 (excluding his 5 per cent commission)[8] but then began using the money to pay his own debts and finance his son's new shipyard at Cardiff. The scheme was to construct a ship there for the Confederate navy and use part of the profit to pay Elswick. But he had to play for time and this required lies and deceit. Russell requested that Elswick use different forms of invoicing and methods of payment, explaining it was at the request of the American government; he claimed that payments were not due until the guns had been delivered to the Liverpool dockside; he let Noble believe that further orders of £100000 were under consideration when, in truth, as time passed and the threat to Boston faded, the Americans were considering ways to cancel the existing orders. By 1865, Armstrong had finished guns in his yard but had not received any money from Russell.

In March 1865, the company finally contacted Adams directly and both parties immediately realized that Russell was withholding the payments. Noble began legal action but Russell played out his bluff to the end, insisting that it was he who had a case against Armstrong for late delivery of goods. But his scheme unravelled when the Civil War ended and his son's shipyard went bankrupt. After a protracted correspondence among lawyers for Russell, Elswick and Massachusetts, Russell finally offered to settle out of court with all his assets: 125 copies of his self-published book, *Naval Architecture*, valued at £42 each.[9] Massachusetts agreed to pay Armstrong £2500, half the sum owed. However, the publisher, to whom Russell also owed money, would not give up the books. Armstrong cried out indignantly: 'A man who cheats you out of £5000 and offers you *a lot of books* as his only assets ... is not exonerated from dishonesty even if you get the books, much less if you don't get them'.[10] In May 1866, Elswick received £2500 from Massachusetts but neither Massachusetts nor Elswick saw any purpose to further legal action.[11] However, the matter did become an issue in Russell's campaign to become president of the Institution of Civil Engineers. The Council of the Institution held an enquiry into the whole matter and expelled Russell.[12] The company wrote off the loss to experience. The American demand for British guns evaporated completely with the end of the war in 1865. The losses from the American arms sale increased Armstrong's caution, but in these early days the risks were very high. Foreign armaments transactions were always open to corrupt agents or governments whose finances were questionable. Armstrong was a very cautious businessman but was willing to take risks to obtain the very high profits that came from the international arms trade. Turkey was such a case, in which the demand for guns was strong but the government's ability or desire to pay made the business risky.

II

The Crimean War had strengthened the bonds between Britain and Turkey and some in Britain looked forward to economic development and liberal reform transforming Turkey into a bulwark against Russian intrusions towards the Mediterranean. A Turkish ironclad fleet was built in British shipyards to replace the one destroyed at Sinope, and Elswick supplied many of its guns. However, modern ships and guns were expensive and Turkey, a non-industrial imperial power in decline, remained on the edge of bankruptcy. Turkey, in short, ordered many Armstrong guns but had difficulty paying for them. In October 1865, a 'large instalment' for guns and £900 for ammunition remained outstanding. Armstrong had 'perfect confidence' in the Turkish government's agent, but he thought that 'it would scarcely be prudent to deliver the guns' until it was confirmed that the money to pay for them was in England. Turkey had also bought expensive ships in England, and Armstrong worried that someone was not going to get paid. He concluded that those 'who can put most pressure on will get a preference', and he asked Rendel to call upon the Turkish ambassador and 'explain these matters to him in the most delicate way'.[13] The ambassador assured Rendel that the money would be forthcoming and the guns were delivered. Armstrong hoped that 'it will all come right'.[14]

The method of payments was given more careful attention in the next contract Rendel negotiated with Turkey, in January 1866. The Turkish demand for a late-delivery penalty clause was balanced by one protecting the firm against late payment, the more likely event in Armstrong's mind.[15] The penalty for late payment was an increase in the time allowed for production of the next batch of guns. This only postponed the problem but it was the best arrangement possible. It was a complex agreement which reflected the productive process. For example, small cannons required less time to manufacture than large ones, but the completion day for each kind was the same in order to maintain an efficient production schedule. For the same reason the completion day for the guns had to merge with the construction timetable of the warships being built elsewhere in England. Armstrong insisted that three equal payments be integrated into this intricate system: a one-third down payment, a one-third payment during production and the final one-third upon completion and before delivery. Armstrong was determined not to repeat previous experience: 'keep the guns until they are paid for [he said, and] all else amounts to little'.[16] The complexities of the negotiations were compounded by the lack of timely discussion among the partners. A disagreement between Armstrong and George Rendel over timing did not emerge until late in the negotiations because Armstrong had not sent Rendel a draft of the contract for his approval because, said Armstrong, 'time was pressing' and he had thought 'everything was understood about times of delivery'.[17] The question was never resolved to Armstrong's satisfaction, but the contract was signed.

Another problem was inspection. Two kinds were possible. The first was a simple demonstration that the finished gun worked. Trials were witnessed by a British officer who attested to each gun's performance. This event had to be incorporated into the contract and made clear, whether or not it was part of the production process. There was no guarantee that an inspector would be available when the guns were finished and this made the time and payment questions debatable. On this point Armstrong told Stuart Rendel, who was conducting the actual negotiations, that if it was necessary to get the contract he might concede the point of extending the time for payment until after the official proof of the gun was completed, provided that the company was 'secured against the risk of delay in getting the Government officer to attend'. He reminded Rendel that the Inspector's Department was very busy.[18]

The second and more rigorous type of inspection involved continual inspection of the guns during construction by a proof officer who would certify that the guns were made according to the specifications in the contract. Armstrong declared that this was unrealistic: 'It is clear', he said, 'that [the officer] will refuse to do this and therefore it is nonsense in inserting it'.[19] The point remained unresolved and Armstrong finally suggested an arbitration clause be added to accommodate any future disputes over interpretation.[20] George Rendel visited Robert Napier, the prominent ship builder who was constructing one of the warships for Turkey,[21] and had what Armstrong called a 'confidential conversation about payments'.[22] Napier was also owed money but he 'seemed to be satisfied that the Turkish Governments were safe though slow'.[23]

Turkey used several tactics to avoid paying for the guns. It demanded that payments be made in Turkey rather than England, but Armstrong saw this as 'only a manoeuvre to enable the Government to dishonour their acceptance without the publicity and damaging effect of doing so in London'.[24] Turkey then accused Elswick of charging Egypt less than Turkey for its guns.[25] Armstrong had sent some guns to Turkey before the final payment had been made and Turkey now refused to pay the £55 000 instalment.[26] Armstrong sent Rendel to the Turkish ambassador with the reminder that 'it is no use his referring us to Constantinople as we get no answer to our letters'. He gave Rendel the name and address of the man who acted on behalf of Napier in these matters in the hope that he would help.[27] As more guns were ready for delivery, Armstrong became willing to compromise for £25 000 but he was determined that 'under no circumstances will we consent *again* to give up possession of goods to the Turkish Government before payment'.[28] If the Turkish government could not meet the payments, the guns might have to be sold elsewhere.[29]

Turkey remained a financially troublesome customer for Elswick over the years, but it seems always to have paid eventually, even though, as we shall see, it required some diplomatic prodding from the British government from time to time.[30] Another early customer, Egypt, was also continually in need of financial relief. In 1878, one year before the final collapse of Egyptian finances, the Chief of Ordnance for the Egyptian government hoped Elswick would 'feel disposed to make such terms as will be advantageous to an old client'.[31] Spain too was constantly in arrears.[32] As Armstrong reminded Rendel in 1887, 'its Agents appear to take a somewhat lax view of their contract obligations'.[33] Despite Armstrong's efforts to maintain the three-payment policy, it often became impractical and had to be suspended.[34] Thus the dangers of selling armaments internationally became apparent early in the company's history: corrupt agents like Russell, and governments like Turkey, whose military needs or ambitions exceeded their ability to pay for modern weapons, were unavoidable. Elswick did its best to accommodate and adapt to these external realities. Fortunately for the company, there were wars which involved countries that could pay and deal with Armstrong with honest agents. One such country was Italy, which established a long and special relationship with Elswick during the war against Austria in 1866.

III

In the midst of Armstrong's worries about Turkish payments, war between Italy and Austria became imminent and a Captain Augustus Albini of the Italian navy arrived in London very anxious to purchase Armstrong naval guns.[35] Albini purchased a large number of 150-pounders in June 1866 and pressed the company for more as quickly as possible. Armstrong hesitated, worried about Turkish legal action if he sold the Turkish guns to Italy.[36] By July, Albini was frantic and even tried to buy the guns of a frigate Turkey had resold (for a good profit) to Chile, which was embroiled in its own war against Spain.[37] Meanwhile, Constantinople anxiously awaited the response from Paris to a loan application.[38] When the loan was refused,[39] the probability that Turkey would pay Elswick diminished further. Italy declared war against Austria on 20 June, only to be promptly defeated at Custozza four days later. The Italian navy's 18 July bombardment of Austrian coastal defences failed and two days later it was defeated at the battle of Lissa. Albini desperately asked for every gun Armstrong could sell him. Armstrong was still concerned about Turkish lawsuits but finally agreed. His hesitations now made time very short. The stunning Prussian victory at Koniggrätz on 3 July left no doubt that the war was rapidly drawing to a close. Armstrong wrote to Rendel that 'Considering the risk that any day may bring us news of the termination of war we do not think it wise to delay selling'.[40]

Albini finally acquired the Turkish guns and made Italy Elswick's most important customer for large naval guns in the 1860s and 1870s. Albini was an excellent ambassador–agent. He became Director General of Arms and Torpedoes at the Italian Naval Ministry, an admiral, a count and an Italian senator, and very active in developing Italy's armaments industry, holding a minority share of the Ansaldo shipyards.[41] Albini was a key promoter of Armstrong naval guns inside the Italian government, a task which required constant attention to the ebb and flow of Italian politics. In 1878, Albini had just succeeded in persuading the minister to purchase new Elswick guns when an impending change of ministry had put the new contract on hold. The change meant Albini would have to begin again with the new minister, 'and you know', he wrote to Rendel, 'how long it takes to persuade a new man upon progress which he has not followed'.[42] The order was further threatened when, on 2 January 1879, a 100-ton gun on the *Thunderer* burst upon firing and strengthened the hands of both the advocates of Krupp guns and those who demanded Italy manufacture her own guns. Albini advised Rendel that George Rendel should come at once to settle technical questions which, even under calmer circumstances, were almost impossible to solve by correspondence, and suggested they reduce the price to speed things along. The contract had to be signed before the *Thunderer* catastrophe became a public scandal 'as no doubt it will grow as soon as the press will [sic] take possession of the question'.[43]

By February, Italian public opinion had been aroused and the Albini–Armstrong faction inside the Italian War Office came under attack, forcing Albini to appoint a committee of inquiry into the disaster. Albini wrote an article to placate opinion with calm words about how the type of gun must be considered 'from a higher point of view', as part of an entire ship, that they were not tied to a particular system and would incorporate new inventions as they came along and proved themselves. He was pleased to report to Rendel that,

> as public opinion demands breechloaders we show that we come to such decisions because a new invention has changed the question and not because the war office demands it. There was no better way to deal with the difficulty of the position and the solution has placed us again in the high ground.

Finally, Albini reminded Rendel of the importance of the similar debate in England because, 'if the public and authorities here see that our ideas are found correct in England they acquire more confidence in our actions'.[44]

The committees of inquiry concluded that the gun had been double-loaded and thus the explosion had not been the result of faulty materials or manufacture. Albini expressed 'great satisfaction' and 'relief of mind' that they had come to the end of a 'laborious battle fought in a very critical time' against powerful adversaries emboldened by the *Thunderer* incident. Final approval by the Council of State was given for the purchase of Elswick guns for two Italian warships. Albini recognised that fresh attacks from the Italian 'Kruppites' and others would continue, but for the moment he enjoyed his victory: 'though very tired of such constant opposition', he said, 'I am not exhausted for fresh battles'.[45] Albini here expressed the same emotional charge that Stuart Rendel had felt during the great Armstrong–Whitworth battle of the guns. Both cases illustrate how deeply politics was embedded in the development of modern industrial armaments.

Economic and political pressures were building up in Italy in the 1880s which affected Elswick's position. Like China and Japan, Italy embarked on a course of industrial self-strengthening based in large part on the development of its own armaments industry.[46] Political independence, modernisation and their own armaments industry were considered inseparable. Unlike China, however, Italy could build an armaments industry, and began to do so in 1884 with the erection of the Terni foundry and steel mill backed by powerful political support and a system of guaranteed orders. In 1882, Italy joined Germany and Austria in the Triple Alliance and committed itself to large increases in military and naval armaments, and Italian foreign policy became bold and contentious towards France while Mediterranean and North African ambitions were offered as compensation for relinquishing irredentism, which had meant poor relations with Austria.

In 1884, Albini warned that Elswick's position was threatened once again. Although the Committee of Artillery had unanimously decided to adopt the Armstrong gun, Elswick could not afford to assume that this decision was permanent. Krupp's allies in the War Office remained a constant threat. Albini pleaded that Elswick 'must make yourselves alive [to the fact that by] remaining in the shade always at distance you are working at a disadvantage with your adversary'.[47] Less than two weeks later, he wrote that 'the state of affairs here has changed a great deal and is changing every day'. Elswick must, he said,

> think serious [*sic*] over the necessity that you should take a decisive step if you do not like to see other people stepping into your field just at the very moment when you were on the point of reaping the benefit of many years [*sic*] battle to reach the result you have obtained.

An Italian engineer, 'whom some people suppose to be backed financially by K...[rupp] or by a French metallurgist' had submitted a proposal to the Italian government to establish a factory to produce steel armour and guns. It was an attractive offer to many, but Albini was confident he could stop it with Armstrong's help. He arranged to have the Italian government offer Armstrong the opportunity to build a factory in Italy which would be guaranteed orders enough to make it profitable for three or four years. Armstrong would be free to import raw materials not available in Italy. Albini had spoken to the relevant minister and was sure that, with such a factory, Armstrong would capture a very large share of the Italian market for both military and naval guns. An Armstrong factory in Italy would, he said, end plans for the Italian government to construct their own or to take up the existing offer. The official letter was on the way, but Albini wrote himself to give Elswick more time to consider the offer and to respond as soon as possible. There was no time to lose as Albini pressed them to answer immediately.[48]

There was little choice really. The Italian market was far too important and was obviously poised for long-term expansion. Italy was an old and important customer which had committed its navy entirely to the Armstrong gun for over a decade.[49] Now the Italian government insisted that future construction be carried out in Italy and it was apparent that a state-subsidised Italian armaments industry was emerging, and thus Armstrong agreed to build a factory at Pozzuoli. George Rendel and Albini took charge of the Pozzuoli Works and in 1888 its first warship was launched.

Although the Pozzuoli Works were profitable, the armaments business there remained intensely political and ferociously competitive. The navy forced Armstrong to buy Italian steel plate from the Terni factory, behind which stood the political figure of Admiral Benedetto Brin, a pioneer of warship design. Naturally, Terni charged Pozzuoli the highest possible prices but, since the ultimate customer was usually the Italian navy, the extra costs were passed back to the Italian state.[50]

The relationship between Pozzuoli and Terni was never friendly and, by the 1900s, their competition had become personal and bitter, and a price war threatened to destroy both of them. John Falkner was sent out from Elswick to attempt to reconcile the two, but he found it hopeless.[51] However, the Pozzuoli factory was profitable as well as necessary, and Elswick indeed expanded its Italian investment in 1904 with the purchase of a share of the Ansaldo Shipbuilding Company of Genoa.[52]

IV

The wars for central Europe in the 1860s drove up the demand for artillery which was powerful, accurate at long range and reliable. However well a new gun proved itself on the testing range – and that was always a matter of debate – performance in war itself remained the final arbitrator of a gun's worth. Yet performance on the battlefield was also a matter of interpretation, and arms manufacturers carefully studied reports on the performance of their guns in battle, advertising their strengths and explaining away any perceived weaknesses. The ultimate aim of arms engineers and manufacturers was to have their design proclaimed the standard. Engineering reputations and business success rested on how performance was interpreted, a complex decision determined by many technological variables and political contexts. Any failure on the battlefield was an opportunity for competitors to gain market share. Such an opportunity arose for Armstrong in the aftermath of the Austro-Prussian War, when the performance of Krupp's cast steel artillery faltered, opening the way for Armstrong's built-up wrought iron guns to penetrate Krupp's monopoly in central Europe. Charles Brackenbury, one of Elswick's most enthusiastic, aggressive and knowledgable agent-ambassadors, believed that such an opportunity presented itself to Armstrong in 1868.

Brackenbury came from an illustrious military family whose roots can be traced back to the fifteenth century.[53] His father and his uncle served in the Napoleonic wars with distinction.[54] His brother, Sir Henry Brackenbury, had an illustrious military career, being promoted to general by Wolseley for his service in the Ashanti War of 1873–4, and chosen as the first Director of Military Intelligence. Charles served in the Crimean War with the horse artillery and went on to a distinguished career in the British army: from artillery instructor at the War Office in 1860, to superintendent of Waltham Abbey gunpowder factory in 1880 and to Director of Artillery Studies at Woolwich in 1886. Brackenbury was an artillery expert who wrote and lectured extensively on the tactical changes required by new weapons. He accompanied the Austrian army and reported on the Austro-Prussian war of 1866 for *The Times* and four years later reported for *The Times* on the Franco-Prussian war. He witnessed the Russo-Turkish War in 1877 (from the Russian side) and commanded an artillery unit in England in 1886–7.[55]

Brackenbury combined the two most important ingredients for a successful armaments agent: thorough technical knowledge of artillery and, as a rising member of the international military elite, access to Europe's military authorities. As Assistant Director of Artillery studies at the War Office in the late 1860s, Brackenbury spent a great deal of time in Germany and Austria examining the developments in artillery and he was thus recruited by Stuart Rendel to act on behalf of Elswick. Thus, like Hart, Brackenbury worked through Rendel in London, not Noble or Armstrong in Newcastle. In 1868, he reported from Berlin that Prussia wanted to make its own guns but was also willing to buy some quietly from outside the country.[56] Elswick prices were attractive compared to Krupp's and Brackenbury was convinced that a great opportunity had presented itself. Indeed, his optimism was unbounded: 'if worked with talent and enterprise, the Armstrong system might be established over the whole of Europe'. He did not mean that all of Europe would buy all its guns from Elswick, but he did believe that at that particular moment the Prussians could be convinced that Armstrong's method of constructing heavy guns was the best. If that happened, orders would *then* come to Elswick. But, he said, they must act quickly.

Brackenbury was convinced that the creation of a European market for the Armstrong gun was possible and he worked hard as an Elswick ambassador in the almost impenetrable markets of Prussia and Austria. He developed friendly relationships with Prussian military officers and politicians who, for their own political–bureaucratic reasons, might be attracted to the Armstrong gun. As Rendel's father had provided access to the Duke of Newcastle in 1854, so too did Brackenbury arrange for Rendel to visit Vienna and Berlin to discuss the acquisition of Armstrong guns.

Brackenbury's enthusiastic ambition for Elswick to seize the opportunity to build support for the Armstrong design in Austria and even Prussia was frustrated by the lack of response from Newcastle. Indeed, the incident suggests a lack of entrepreneurial vision by Armstrong and his partners. Explaining the need for immediate, quiet political action by 'an envoy with full powers' in Berlin, Brackenbury pleaded to Rendel:

> It is no use my writing to Elswick. I know the sort of arrogant stupidity with which I should be answered. I will tell you more as things progress with me here. At present if I ruled at Elswick I should ask you to come to Berlin with instructions to do what you liked as long as you worked to secure the influence of the firm over the Prussian Government.[57]

But Brackenbury reluctantly recognised that it was not his business to do more than point out general opportunities to the firm. He saw similar opportunities in Austria, but there too he became frustrated with the apparent lack of interest at Elswick:

> Have your people at Elswick sense enough to put the making of this muzzle loader for Austria before everything? Do they see that by the end of the next year matters may be settled, and Europe may disarm more or less? First Armageddon and then the golden age – or the golden age and then Armageddon. Which?[58]

Brackenbury believed the former, and he worried 'that Elswick should let the grass grow under its feet'. The military balance in Eastern Europe had been upset by the Austro-Prussian War: all that stood between Russia and Turkey was Austria's old bronze artillery. Armstrong guns, he believed, could make a difference.

Krupp and Prussian national security considerations were formidable obstacles however. As Siemens told Brackenbury, 'we won't have any outland manufacturers in Prussia and that is one great reason why we reject the Armstrong gun'.[59] But it was not quite that simple. British and European military officers often saw no reason to support their home armaments industry and, indeed, had good reason to buy from beyond their own lands, openly if possible, secretly if necessary. In 1863, Sir William had suffered the humiliation of seeing Alfred Krupp welcomed in London by the British military. No guns were purchased, but some Krupp steel barrels were bought secretly.[60] Before the Austro-Prussian War, Krupp had been able to stop the purchase of Armstrong's muzzle-loaders by the Prussian military who, like their British cousins, had little compunction about such foreign purchases. A visit to Bismarck in May 1862 fended off this Armstrong threat.[61] After the war, however, the Prussian military considered a return to muzzle-loaders, and the best of those were made at Elswick, not Essen. Armstrong's opportunity to enter Krupp's kingdom came in the later 1860s. While the breech-loading rifles had been a tremendous success against the Austrians in 1866, Krupp's breech-loaded cast steel artillery had been a disaster. Guns burst, killing many Prussian gunners. Krupp's reputation suffered a severe blow and he fled into self-imposed exile in France for a year.[62]

The poor performance of Krupp's guns precipitated a political struggle inside the Prussian military between those who had always supported Krupp and those who wanted to free themselves from a Krupp monopoly. Because the technical failure of Krupp's guns was a matter of interpretation tied to Prussian politics, Brackenbury was quite correct to conceptualise the opportunity as more political than technological. The anti-Krupp faction in the Prussian War Office sought to replace Krupp with a government armament factory, but until such facilities could be built they needed to buy guns. It was this situation which gave Elswick its opportunity. As Brackenbury wrote to Armstrong, the Prussian Artillery Committee were 'at their wits [*sic*] ends to find means of making heavy guns outside Krupp's factory' and, if Armstrong 'would work quietly and privately', he might gain converts to his gun inside the Prussian military. A competition between an Armstrong and a Krupp gun was planned, but Brackenbury argued that 'little or nothing is to be done by official letters or trials'.[63] The opportunity was political, not technological.

Recognising the political nature of the technological question, Brackenbury advised caution. Aggressive tactics should be avoided, he said, and certainly there should be no public debate like the one which had taken place between Armstrong and Whitworth ten years earlier. Against Krupp in Germany, such a strategy would not work. A delicate balance had to be maintained between the international armaments business, German government goals and nationalist sentiments. The

Prussian government was sensitive to the rising costs of armaments, while the Prussian military struggled against the domination of Krupp. The Government was 'sharp enough to keep up the credit of their own guns and manufacturers, but also sharp enough to take advantage of any quiet opening for adopting the best and cheapest system they can find'. With quiet diplomacy and negotiations shaped to bring long-term market share benefits to Elswick rather than immediate profits, Brackenbury was convinced that the Prussian government would adopt the Armstrong system. To this end, he told Rendel, 'it would be necessary to yield a point here and there[64] and to endeavour to introduce your system whether you had orders at Elswick or not'.[65] In short, Brackenbury advocated that Elswick negotiate with Germany in 1868–9 an arrangement which would make the Armstrong gun the standard design in Prussia. If this was even partially achieved the Armstrong brand would be enhanced in the same way as it had been in 1859–69 when the British navy adopted the first Armstrong guns. Acceptance of the Armstrong design by Prussia – even without Prussia buying any guns from Elswick – would establish Armstrong's design as the world standard. Given the interpretive nature of technology, whether or not the Armstrong wrought-iron built up design was 'superior' to Krupp's cast steel cannon could not be quantified objectively and was hence irrelevant. The Prussian political and institutional complexities, however, could be measured.

After witnessing trials between guns from the two companies, Brackenbury recommended that Elswick open itself to learn from Krupp. For example, in his opinion, Krupp's prismatic powder made 'an enormous difference' and he suggested that Armstrong adopt it at once: 'don't be too proud to learn from them', he said, 'as they have learned carriages, shape of shot, etc. from you'. Brackenbury acknowledged that the struggle with Krupp would be severe, but Elswick 'must not throw away a chance' by ignoring the technological advances made by Krupp. The 'Kruppites' were clearly worried about how they were doing in the competition.[66] One important advantage for Armstrong was that the Prussian officers were 'abominably ignorant about the *experimental* science of artillery'.[67] This ignorance intensified their dilemma over the Krupp monopoly and made them more susceptible to an offer from Armstrong. Thus disarray among the military–political factions in Germany presented an opportunity for Elswick to gain a foothold in the Prussian market. One Admiral von Tegethoff was an advocate of breech loading but not a Kruppite and, after a long talk with Brackenbury, agreed that the Berlin trials 'were somewhat of a swindle', and further agreed to help find an officer to act on behalf of Elswick. Tegethoff had his own motives: he was, said Brackenbury, 'a member of His Prussian

Majesty's opposition and one who would be glad to come in again on any measure'.[68] But Brackenbury reiterated the crucial necessity of quiet diplomacy:

> Your naval friends [he explained to Rendel] made a great blunder. They claimed too great a superiority for your guns in every particular. If the navy had been content to say 'whether the guns be equal or the Krupp even a little superior the A. suits us best,' they might have had the muzzle loaders without much trouble but the case was put as if there was an over-whelming superiority of British guns over German. There lay the sting. The attitude of the Government to the people cannot permit them to acknowledge any great blunder or want of warlike skill. They are obliged to surround themselves with an atmosphere of watchful deception.[69]

There was another factor to be considered. The Germans had taken the Armstrong gun-carriage system 'without reward or thanks', a common practice by governments in a world without international patent agreements to control it. Indeed, Josiah Vavasseur, whose carriage technology gained him a seat on the board of Armstrong's company in 1883, had learned in 1878 that British courts placed the rights of sovereigns above the property rights of individuals in the patent laws.[70] But if the German military was persuaded to do business with Armstrong, Brackenbury's discussions with the Berlin War Office indicated that some recompense might be obtained.[71]

In late 1869, then, opportunities in Prussia and Austria were opening up and Elswick had a man in place with both the enthusiasm and the ability to take advantage of the new situation. But, as with other agents and ambassadors, Brackenbury's work on behalf of Elswick was not his only concern. He was in Europe on leave from the army, and he made sharp distinctions between his roles as an officer in the British army, a correspondent for *The Times* and an independent advocate of the Sir William G. Armstrong & Company (on behalf of his friend, Stuart Rendel). His European tour began in Vienna where he openly advocated the Armstrong gun to Austria and accordingly requested Elswick pay for his accommodation.[72] But in Berlin the case was different. He went there as a reporter: 'my business is now with *The Times* and I must take care not to mix myself up with any manufacturing interests'.[73] Certainly, Brackenbury practised what he preached: his articles for *The Times* on the Berlin trials were cautious and discreet. His reports were written for two audiences – the British public and the Berlin War Office – and he attempted to appeal to both the general reader and the experts. With the latter in mind, he 'really tried hard to be honest and not to overstate any facts or arguments'.[74]

In Austria too a way had opened for Armstrong to expand his business. Austrian military confidence had suffered a sharp decline after the Seven Weeks War and was open to anyone who promised help. Brackenbury claimed that the Austrian armaments factory of Joseph Wernal was available for purchase cheap and he pushed hard to have Armstrong buy it. In his optimistic opinion, 'a man of real insight, talent and energy might make Wernal's factory into an Austrian Krupp's'. Machinery would have to be brought in, but, like Krupp's, the Wernal works 'were built in quite an extravagant style'.[75] All that the factory really needed was 'English honesty and management and that,' he argued, 'would have been answered by Sir William's connection with it'.[76] If Elswick would 'send good men out to work the factory and financial affairs', Brackenbury felt there was 'no doubt of every help being given by the Austrian Government'.[77] It is highly doubtful that Wernal's was for sale, and most certainly not to foreigners. Wernal was well-connected with the Austrian military, both socially by marrying his daughters to artillery officers, and technologically with his innovations in small arms.[78]

Brackenbury was too enthusiastic about the project and overly optimistic about the future role of England in Austria and, in cooperation with Austria, in Turkey.[79] He pressed Rendel to push Elswick to decide quickly, threatening to act on behalf of Whitworth if Elswick declined, insisting that 'such a chance of another bond between England and Austria is not to be lost'.[80] When he learned that the firm would not entertain the idea of working the Austrian factory, Brackenbury withdrew from the entire affair[81] and returned to England. Brackenbury travelled often in Europe, however, and he kept himself available to help his friend, Stuart Rendel.[82] But decisions were made in Newcastle, and nothing was done. When Krupp rehabilitated himself and his artillery in the Franco-Prussian War, the opportunity passed.

The story above seems to support the case against British entrepreneurs for failing to market their goods effectively. Armstrong appears reluctant to seize the momentary opportunity of a lifetime to capture a good slice of the nearby Continental European market in field artillery, a market which was possibly larger than the rest of the world combined, including Britain. Is this episode not a sign of a long-term underlying entrepreneurial weakness in marketing by Armstrong? The huge Continental armies of Prussia and Austria were surely a market which Elswick should have pursued with much greater vigour in the aftermath of Krupp's 1866 troubles. Was this not an opportunity passed over?

It is an appropriate question, but one to which only ambiguous answers can be given, in part owing to the complexities of the interpretive nature of technological performance and the political context in which a particular piece of technology is accepted as superior to another. Brackenbury provided Elswick with a brilliant and no doubt accurate analysis of the situation. But he too was unsure when the core of the matter was 'purely' technological and when it was political. Brackenbury's bitter complaints in October 1868 about Elswick's arrogance, stupidity and slowness are reversed in December, when he advises against aggressive salesmanship and cautions Rendel to restrain Elswick's naval friends from making too great a claim about the qualities of Armstrong guns over those of Krupp, for fear of stimulating a psychological reactionary defence mechanism against the English gun. Moreover, trials of Krupp guns in December made Brackenbury realise that Krupp technology was not the disaster it had seemed earlier; indeed, he now wrote about Krupp teaching Armstrong a thing or two about artillery.

Brackenbury's shifting views only confirm that the interesting question is not whether or not Armstrong stumbled as an entrepreneur in 1868, but centres on the interpretive and political nature of technological design. As with the competition between Armstrong and Whitworth in 1863–4, the debate over ordnance design in Prussia in 1868 was also a political struggle over defining what exactly constituted the best artillery technology: what were the best materials and design – which of Krupp's cast steel or Armstrong's wrought iron designs produced the specific effects deemed most efficient.

Armstrong had won this debate over his fellow countryman in 1858 and again in 1864, but his political and bureaucratic support in Prussia was too weak for him to win that battle against Krupp in 1868. Whatever the technological problems of Krupp's gun, and despite those who saw them as an opportunity to break Krupp's monopoly over Prussian ordnance (as Woolwich defended its monopoly of British ordnance), Krupp enjoyed unswerving support from the top of Prussia's military bureaucracy and political leadership, from General Constantin von Voigts-Rhetz, who was a strong supporter of steel cannon and had been responsible for Krupp breaking into the Prussian market, and the far more weighty support from the King of Prussia, William I, a personal friend of Alfred Krupp.[83] Armstrong 'failed' to gain a foothold in Prussia because William I remained attached to Krupp's cast steel barrels.[84] This doomed the anti-Krupp faction in the Prussian military from the start and, as Brackenbury recognised, Armstrong's foothold in Prussia depended entirely on their political success. That was never in doubt. Bismarck too agreed with Krupp that Armstrong must be kept out.

Krupp thus enjoyed something the equivalent of which Armstrong never possessed: the unbreakable bond between himself and his King and later Emperor. In 1863, the anti-Armstrong faction inside the War Office and in Parliament defeated Armstrong because he lost all his political support. Such connections were absolutely essential to success in the armaments business. It was Krupp's great advantage over Armstrong in that, unlike Armstrong, Krupp was, from the beginning, a private firm which was treated as a special branch of the state. In Britain, the private production of armaments was always looked upon as a secondary, additional component for emergencies only, sacrificed to Woolwich in times of calm. By 1880, the Krupp firm was the largest armaments manufacturer in the world; it continued as Germany's source of ordnance through defeat in two world wars, and became a legend. Does this make Krupp the greatest entrepreneur of his age and, if so, was it because of his technology, his marketing or because he was given an absolute monopoly by the German state? He was bigger than Armstrong and all other firms, but can the entrepreneurial achievements of the various armaments firms be comparatively measured without some reference to the quite different political contexts in which each existed? When the political situation in Britain changed in the 1880s, Armstrong and Elswick reaped the rewards. These fluctuating conjunctional factors cannot be incorporated into economic models of entrepreneurial performance. The whole question of British foreign marketing entrepreneurship is extremely complex and, as one historian has rightly observed, 'it is remarkable that the judgement of bad salesmanship should have been delivered without more empirical research'.[85] Perhaps it is also time for historians to ask whether or not the law of diminishing returns has taken its toll on the debate over the performance of British Victorian entrepreneurs, and to ask whether the time had come to add some political and contextual questions to the debate in order to increase and expand our understanding of these matters.

The Franco-Prussian War of 1870–71 was the last of a series of large European and American wars which had begun with the Crimean War of 1854–6 and had been made extraordinarily murderous by the new rifles and artillery developed in the 1840s and 1850s, and improved during the 1860s. War re-established the American nation-state, forged the new nation-states of Italy and Germany and exhausted the South American republics. War erupted from time to time after 1871 – in Turkey in 1877 and in South America in 1879 – but generally the 1870s was a decade of relative peace for most of the world and the frenzy of demand for armaments which had characterised the 1860s subsided. In Asia, however, where the demand for modern arms during the 1860s had been relatively small, the 1870s proved to be a turning point in the demand for guns and ships as China and Japan stepped forward as new and important customers for Elswick.

Notes

1. For overviews of the industrialisation and commercialisation of armaments and the changing nature of warfare which began in the 1860s, see Theodore Ropp, *War in the Modern World* (New York: Collier Books, 1962), ch. 6; Brian Bond, *War and Society in Europe, 1870–1970* (London: Fontana, 1984), ch. 1; Paul Kennedy, *The Rise and Fall of the Great Powers* (London: Fontana, 1988), ch. 4; William H. McNeill, *The Pursuit of Power: Technology, Armed Force and Society Since A.D. 1000* (Chicago: University of Chicago Press, 1982), ch. 7.
2. Alfred D. Chandler, *The Visible Hand: The Managerial Revolution in American Business* (Cambridge, Mass.: Harvard University Press, 1977), pp. 51, 64, 72–7, 271, 308.
3. *The Economist*, vol. 19 (7 Dec. 1861), p. 1350.
4. *The Times*, 9 January 1862; 'Royal Commission on Warlike Stores', *Parl. Papers* XV (1887), Evidence q. 7521. Armstrong denied he sold guns to the Confederate government. See *The Times*, 21 January 1864, p. 9 and 16 June 1864, p. 11.
5. George Emmerson, *John Scott Russell: A Great Victorian Engineer and Naval Architect* (London: John Murray, 1977), p. 190.
6. Institution of Civil Engineers, *Sir W.G. Armstrong and Mr. J. Scott Russell: Correspondence and Documents Submitted to, with Minutes of Evidence Taken Before, the Council of the Institution of Civil Engineers* (London, 1867), p. 195.
7. Adams to Rendel, 21 January 1864, T&W, 31/204. The £5000 would buy one 600-pounder and several hundred rounds of ammunition. Other size guns were also ordered. See Institution of Civil Engineers, *Armstrong and Russell*, pp. 70, 72–3.
8. T&W, 31/205.
9. J. Scott Russell, *Naval Architecture* (1865).
10. T&W, 31/209.
11. Adams to Rendel, T&W, 31/204.
12. The inquiry is discussed at length by Emmerson, *John Scott Russell*, but his defence of Russell ignores several vital pieces of evidence (including Russell's confession) and distorts others. His arguments and conclusions must be checked against the Institution of Civil Engineers, *Armstrong and Russell* and L.T.C. Rolt, *Isambard Kingdom Brunel* (London: Green, 1957).
13. T&W, 31/106.
14. T&W, 31/108.
15. T&W, 31/134 and /135. Late delivery penalties in 1877 were 5 per cent of the price for the first month and 2½ per cent for each succeeding month. See Radcliffe, Cator and Martineau to Derby, 3 May 1877, F.O. 78/2664.
16. T&W, 31/133.
17. T&W, 31/140.
18. T&W, 31/132.
19. T&W, 31/133.
20. T&W, 31/137.
21. T&W, 31/149; Napier biography, *D.N.B.*
22. T&W, 31/148.
23. Ibid.

24. T&W, 31/149.
25. T&W, 31/104.
26. T&W, 31/3136.
27. T&W, 31/149.
28. T&W, 31/154.
29. Armstrong to Rendel, T&W, 31/149.
30. See Chapter 9.
31. T&W, 31/5264.
32. T&W, 31/2977.
33. Ibid.
34. T&W, 31/44, /2977.
35. T&W, 31/152, /153, /160 and /163; Rendel, *Personal Papers*, p. 278.
36. T&W, 31/152, /157, /158, /160 and /164.
37. T&W, 31/163, /164.
38. Armstrong thought it was for £2 million; T&W, 31/164.
39. *The Times*, 10 July 1866, p. 10.
40. T&W, 31/167 (18 July 1866).
41. R.A. Webster, *Industrial Imperialism in Italy 1908–1915* (Berkeley, Ca.: University of California Press, 1975), p. 81.
42. Albini to Rendel, 27 Nov. 1878, T&W, 31/4641.
43. Albini to Rendel, 6 Jan. 1879, T&W, 31/4643.
44. Albini to Rendel, 6 Feb. 1879, T&W, 31/4645.
45. Albini to Rendel, 3 June 1879, T&W, 31/4651.
46. Webster, *Industrial Imperialism*.
47. Albini to Rendel, 5 April 1884, T&W, 31/1932.
48. Albini to Rendel, 17 April 1884, T&W, 31/3110.
49. *The Times*, 12 Feb. 1868, p. 7; 7 Sept. 1868, p. 10.
50. Webster, *Industrial Imperialism*, pp. 54–6.
51. Falkner to Rendel, 13 Jan. 1907, T&W, 31/7057.
52. T&W, 31/3654. The company had been discussing such a move for years. See Falkner to Rendel, 27 Feb. 1900, T&W, 31/7040.
53. Sir Robert Brackenbury, *D.N.B.* (vol. 22).
54. Ibid.
55. Ibid. Cardwell's 1870 army reforms banned officers writing for newspapers. See *The History of The Times*, vol. 2 (London, 1939), p. 420.
56. Brackenbury to Rendel (27 Oct. 1868), T&W, 31/1033.
57. Ibid. Two months later, Brackenbury wrote to urge that Rendel be sent to Berlin (Brackenbury to Armstrong, 30 Dec. 1868, T&W, 31/1041).
58. Ibid.
59. Brackenbury to Rendel, 29 Dec. 1868, T&W, 31/1040.
60. William Manchester, *The Arms of Krupp 1587–1968* (New York: Bantam Books, 1970), p. 104.
61. Ibid. The 1859 contract between Armstrong and the British government covered only breech-loaders. He was always free to sell muzzle-loaders in foreign markets.
62. Ibid., pp. 111–13.
63. Brackenbury to Armstrong, 30 Dec. 1868, T&W, 31/1041.
64. Brackenbury probably meant price and credit, but also, and in his mind more importantly, the construction of a factory in Germany.
65. Brackenbury to Rendel, 27 Oct. 1868, T&W, 31/1033.
66. Brackenbury to Rendel, 29 Dec. 1868, T&W, 31/1040.

67. Ibid.
68. Ibid.
69. Brackenbury to Rendel, 1 Nov. 1869, T&W, 31/1045.
70. *Vavasseur* v. *Krupp*, *The Law Reports Chancery Division*, vol. 9 (1878), pp. 351–62.
71. Brackenbury to Rendel, 1 Nov. 1869, T&W, 31/1045.
72. As he explained to Rendel, 'it is an expensive place especially for me because I have to keep up a certain appearance' (T&W, 31/1034, 3 Nov. 1868).
73. Brackenbury to Armstrong, 30 Dec. 1868, T&W, 31/1041. Brackenbury was quite sincere about these distinctions. See T&W, 31/1037.
74. Brackenbury to Rendel, 28 Jan. 1869, T&W, 31/1046. See also 31/1037.
75. Brackenbury to Rendel, 29 April 1869, T&W, 31/1051. 'Most continental manufacturers,' he added, 'take out in building what they save by cheap labour.'
76. Ibid.
77. Brackenbury to Rendel, 29 April 1869, T&W, 31/1050.
78. Andrew Wheatcroft, 'Technology and the Military Mind: Austria 1866–1914', in Geoffrey Best and Andrew Wheatcroft (eds), *War, Economy and the Military Mind* (London: Croom Helm, 1976), p. 53.
79. 'The Eastern Question,' he said, 'is drawing near to a solution in better form than was to be hoped' (ibid.).
80. Ibid.
81. Brackenbury to Rendel, 29 April 1869, T&W, 31/1051.
82. Ibid.
83. Manchester, *Arms of Krupp*, pp. 94–9, 111–16, 120–21, 139–40, 154–7, 191–3.
84. These political obstacles also frustrated the efforts of the Prussian military to develop their own bronze gun and thereby break the Krupp monopoly. See Brackenbury to Rendel, 29 Dec. 1868, T&W, 31/1040.
85. Stephen J. Nicholas, 'The Overseas Marketing Performance of British Industry, 1870–1914', *Economic History Review*, 2nd ser. 37 (1984), p. 490.

CHAPTER SIX

Arming Asia in the age of imperialism

I

Before 1500, China was the most technologically innovative country in the world: gunpowder was invented in the ninth century, printing in the eleventh, ironclad warships in the twelfth, projectile-throwing devices in the thirteenth, and gunpowder artillery and compass-guided long-range ships in the fourteenth century made the Mingh dynasty the towering imperialist power in Asia. A maritime nation since the eleventh century, China sent out a number of naval expeditions in the fifteenth century, five decades before Portugal launched the European age of discovery. Had not Confucian intellectuals at court persuaded the emperor that world exploration was a sign of decline, China might well have discovered Europe at a time when European armies still brandished their swords, lances and bows, while its navies went about ramming and boarding one another.[1]

If technology is the autonomous process that the internalist and technological pessimists think, then China should have taken off with a scientific–industrial–commercial revolution in the fifteenth and sixteenth centuries, sailed great battleships carrying huge cannons into European harbours and sent steam-powered gunboats up its rivers to bombard Lisbon, Bordeaux, Paris and London into submitting to the Chinese world order. Yet it was Europe that sailed to China and the rest of the world in the sixteenth century; it was Europe, not China, that launched a great military revolution of guns and ocean-sailing warships, with which they increasingly exploited the peoples and natural wealth of the world, including those of China; and Europe which developed and refined gunpowder technology during the seventeenth and eighteenth centuries until, in the nineteenth century, the great states of Europe acquired the industrial weapons with which they imposed their increasingly overpowering domination of the world, and nowhere more overwhelmingly and ruthlessly applied than in China, in the age of imperialism.[2]

That the global military balance of power had shifted in favour of the west was recognized in China as early as 1517, when an official warned that western artillery was far more advanced: 'No weapon ever made since memorable antiquity is superior to their cannon.'[3] That Europe and not China rose to world dominance after 1500 is evidence that technological innovation is closely related to and shaped by the political and cultural context in which it occurs. Whereas China's absolutist Confucian state successfully resisted technological and commercial industrial evolution as a threat to its power, early modern European governments embraced technology as a means to increase and expand their absolutist powers.

European states encouraged science and technology in order to increase their military power, but also because science and technology was a part of western culture, and it was natural for kings, emperors, prime ministers and lowly engineers like Sir William Armstrong to link the two together. The scientific revolution of the seventeenth century and the Industrial Revolution of the late eighteenth and nineteenth centuries comprise the dual revolution which enriched European states, but it was not technologically inevitable that Europe would also turn itself into a world naval and military power. As one historian who writes of the relationship between economic and military power has rightly observed, 'an economic giant could prefer, for reasons of its political culture or geographic security, to be a military pigmy, while a state without great economic resources could nonetheless so organize its society as to be a formidable military power'.[4] France and Russia are examples of states which sought to regain military supremacy in the nineteenth century through technological innovation, but failed for political and cultural reasons. Germany achieved great industrial power in the late nineteenth century but found that military and naval power was not in itself enough to achieve its political ends.

The relationship between industrial capacity, military strength and political culture is perhaps most complex in nineteenth-century China, which became an important market for Elswick naval weapons in 1874, when the Chinese leaders realised that their social and political order might be toppled either by rebellion from within or by the persistent intrusions of the world's senior great powers, Britain, France and Russia, and of two new naval powers, Japan and the United States.

Throughout the nineteenth century, Britain led the European great powers in pressing China to join their global economic and political system, refusing to accept China's rejection of European diplomatic and commercial relations in 1793, 1816 and 1834. When China cut off opium imports in the 1830s, British patience evaporated and turned China's four hundred year-old gift of gunpowder against its maker.[5] The Opium Wars of 1839–42, the *Arrow* war of 1856–8 and the British and French invasion of 1860 made many Chinese leaders realise that their centuries-old ingrained sense of self-sufficiency and superiority over foreigners did not extend to military and naval matters. This lesson was underlined by the inability of the Imperial government to crush the Taiping Rebellion which swept across China between 1850 and 1864, the Nien Rebellion between 1852 and 1868, and the Muslim rebellion in the north east from 1862 to 1875 (the latter supported by arms from the British government in India, which would have welcomed a strong Muslim state in the north of China as a barrier to Russian incursions towards India).[6]

In 1861, China responded to its weak position by launching a campaign to strengthen the existing order against internal rebellion on the one hand and foreign encroachments on the other by acquiring western industrial technology and armaments. This great 'self-strengthening' movement required China to accept closer diplomatic and commercial relations with the western industrial powers. A new office of foreign affairs, the Tsungli Yamen, and an Imperial Maritime Customs office were established in 1861. Rutherford Alcock, a British diplomat with many years experience in China, called these changes 'such a revolution as has never been seen since the world began'.[7]

It was certainly a major turning point in China's history, and must be counted among the first significant episodes in the age of imperialism and modern global history. The internal upheavals and external intrusions in China in the second half of the nineteenth century brought the 'self-strengthening' reformers to the political forefront in China's imperial state. There were many self-strengtheners, but four stand out: two Chinese, Tseng Kuo-fan and Li Hung-chang, and two British Inspectors General of the Chinese Maritime Customs Service, Horatio Nelson Lay and his successor Sir Robert Hart, through whom Armstrong would obtain his first orders for gunboats.[8] Next to the land tax, the Maritime Customs Service was the most important source of income with which China could build and buy armaments.[9]

Tseng and Li were provincial civilian leaders who were more responsible than any of the Chinese imperial military authorities for defeating the Taiping armies. Tseng's personal Hunan army had been far more effective than the regular imperial army and his final victory over the Taiping kingdom at Nanking in 1864 established him as a powerful force in Chinese imperial politics. Li was a distinguished scholar without office in 1853 when the Taipings attacked his home city in Anhwei. He raised a militia which grew under his energetic and able leadership into an army of between twenty-five and forty thousand. He soon attracted the attention of Tseng Kuo-fan who accelerated his political career. Both recognized the need for western guns and ships, the superiority of which they saw demonstrated during the invasions carried out by Britain and France between 1856 and 1860.[10] Tseng actively sought the secrets of

European gun making and ship building with which he could crush the Taiping rebellion, on the one hand, and counter the naval and military might of the foreign powers, on the other. He encouraged the study of western military techniques and in 1865 established the Kiangnan Arsenal in Shanghai to manufacture guns. By 1867, a tenth of the Shanghai Customs revenue was allotted to him to construct European-type gunboats.

Tseng Kuo-fan died in 1872 and the leadership of the self-strengthening movement passed to Li Hung-chang. Li had visited British and French warships in 1863 and came away amazed by their ordnance and was moved to write that he was 'deeply ashamed that Chinese weapons [were] far inferior to those of foreign countries'.[11] Like Tseng, Li became convinced that western weapons would provide China with security from internal and external threats and he became the architect of China's military and naval reconstruction and its principal representative abroad. Li promoted and built arsenals and struggled to reorganise China's army on the Prussian model and her navy on the British model. He built a small gun factory in Soochow in 1864, the Nanking Arsenal in 1867, expanded a machine factory in Tientsin in 1870, sent army officers to study in Germany, naval officers to study in Britain and organised naval and military academies at Tientsin to support his ambitious plans for a modern Chinese navy. Li extended Tseng's concept of military and naval self-strengthening to China's economy, promoting industrialisation, building machine factories, steamship companies, textile mills, mines and railroads.[12] Li Hung Chang was behind virtually every project aimed at strengthening China to withstand the threats which grew more dangerous as the century entered its last decades. Although officially he was only the governor-general of Chihili province, Li's political power in the central government exceeded that of the Tsungli Yamen. He negotiated foreign treaties (with Japan in 1895, for example) and personally visited Britain, Europe and the United States to buy armaments and meet the leaders of the great powers.

II

China's first attempt to purchase a navy failed. In 1861, the Taipings were still a threat to the imperial government and the Tsungli Yamen and Tseng were convinced by Hart, who was acting Inspector General while Lay was in England on leave, that they should purchase a small fleet of steam gunboats which could be used to combat smugglers after the rebellion was put down.[13] With Hart peppering him with letters from China to act with haste, Lay purchased eight gunboats and negotiated with the Admiralty and the Foreign Office for the services of Captain Sherard Osborne to command the little fleet and its crew of British naval officers and men. What became known as the Lay–Osborne fleet arrived in China in 1863, whereupon the entire project was scuttled by Lay's determination to exercise complete control over the ships. The Tsungli Yamen and Tseng Kuo-fan insisted it come under Chinese political authority. Lay would not give way and the fleet was returned to England.[14] The Taipings were defeated in 1864 and, with better relations with the west, a Muslim rebellion and Russian pressure in the north, the incentive to build a navy faded.[15]

The second attempt by China to acquire a small navy was much more successful. China was reminded of its need for a naval force in 1874, when it was forced to accede to the demands of a stronger foreigner. This time it was not a western power which humiliated China but her eastern neighbour, Japan. In 1871, 54 sailors from the Ryukuan Islands, which lie between Japan and Korea, were shipwrecked off Formosa and killed by Formosa aborigines. The Ryukyuans, once tributary to China, were now dominated by Japan and, three years after the killings, Japan invaded Formosa, an action that was a compromise between those who demanded an expedition be sent to Korea and those who were not so bold. China could not defend its position and paid Japan a £2 million indemnity for the murdered sailors and for the costs of the expedition, a concession which recognised Japan's claim over the islands.[16]

The crisis renewed the pressure on China to obtain at least a small naval force to defend its coastal waters. Among those who wanted to build a modern Chinese navy was Sir Robert Hart, who had succeeded Lay as Inspector General of the Chinese Imperial Maritime Customs Service. In the midst of the Formosa crisis of 1874, an Armstrong agent by the name of Bidwell arrived in China and introduced himself to Hart:

> B. could not arrive in China at a better time [he wrote to his London representative]. This Formosa affair has given everything a little fillip, and the Mandarins are ready to try iron-clads, monitors, catapults, cross-bows, Krupps, *old c/o's* – and anything else that anybody will tell them to acquire.[17]

Bidwell no doubt discussed with Hart the idea of purchasing coastal warships which would support an expanded Imperial Customs Service and Hart convinced Li to allow him to purchase some gunboats. He thus instructed his London representative, James D. Campbell, '*to ascertain most quietly and most confidentially*' the cost of 'a most-peaceable-looking steamer' but which made 15 knots and carried a single, fixed and hidden gun which could penetrate 20 inches of armour at 500 yards, and stressed that he did not want '*any one to know that China contemplates anything of the kind*'.[18] Neither the *Staunch* nor Elswick was mentioned, but Hart's detailed specifications of the vessel he wanted suggests that Bidwell had recommended that gunboat. Campbell knew nothing of armaments and sought advice from the Admiralty who directed him to Elswick and the *Staunch*.[19] Campbell later claimed the credit for bringing the Chinese orders to Elswick: 'if I had not been here,' he wrote to Stuart Rendel, 'you would never have had the order probably for these gunboats'.[20] Rendel claimed he had changed the original Chinese demand for ironclads (which Elswick did not build at the time) to 'mere defensive gun vessels' designed by his brother (which Elswick did make). Hart, he said, 'had no leaning towards business of this character for China'.[21] The evidence is inconclusive on this point but Campbell, who never hesitated to promote his own importance in these matters, recognised Rendel's important role:

> I consider [he told Rendel] that the kind assistance and advice which you gave me during the execution of the order for gunboats materially contributed to the success achieved, whilst the communications I have had with you from time to time have enabled me to send Mr. Hart information that would be useful to him in advising the Chinese authorities.[22]

Negotiations were cumbersome as they had to be done by telegrams and letters. Hart pressed his case on the Chinese and communicated with Campbell, who negotiated the contract in England. During the Formosa crisis, Campbell received a flurry of telegrams from Hart to obtain estimates for armaments. But Campbell's position in England was also uncertain. The purchase of arms and gunboats and the recruitment of officers to operate them required negotiations with the Admiralty, the War Office and the Foreign Office. China did not have ministerial representation in England until 1876 and Hart did not want to lose his control over his London agent Campbell by having Campbell take on a consular rank in London. Campbell was therefore given only an informal recognition by the British Legation in China.[23] The main negotiators were Rendel and Hart; the two men exchanged many letters and a close personal relationship developed between them. Hart relied upon Rendel and trusted him completely, even to the point of seeing no need to have an outside inspector at Elswick during construction of the gunboats.[24]

Sir Robert Hart was one of Elswick's most productive foreign ambassadors. Chinese orders from Hart provided the company with invaluable experience building new and advanced warships, leading up to the new cruisers built in the 1880s. China provided a steady market for Elswick's *Staunch* gunboats, which quickly developed into formidable coastal warships that relatively poor countries could afford. Four *Staunch* gunboats were ordered for China in 1875 and over the next nine years China purchased more than a dozen gunboats and two cruisers from Elswick.[25] The first two *Staunch* class gunboats for China, named *Alpha* and *Beta*, were 330 tons, of 7 1/2 inches of draught, unarmoured (except around the gun) and had a single engine which delivered 9 knots. It carried a single 26-ton Armstrong rifled muzzle-loaded gun at the bow. It took five months to build and cost, with armament and 100 rounds of ammunition, £23 000. The second and third boats, *Gamma* and *Delta*, were each 440 tons, carried a 38-ton gun and a 12-pounder breech-loader, were delivered in nine months and cost £33 400. Newer boats were progressively more powerful. The *Epsilon* group (*Epsilon*, *Zeta*, *Eta* and *Theta*) were also 440 tons but their two compound engines and twin screws delivered 10 knots, the hulls were steel instead of iron and they were divided into four water-tight compartments. They carried the new Armstrong 11-inch, 35-ton gun, (far more powerful than the earlier 38-ton gun) which had a range of 7500 yards and required only five men to work it. Each also carried two 12-pounder breech-loaders and the new ten-chambered Gatling machine gun, for which Armstrong held the licence to manufacture in Britain. In 1880, China purchased two Armstrong cruisers which carried two heavy guns and spar torpedoes.[26] By 1880, China had a small but modern navy with the latest technology from the world's foremost armaments manufacturer.

Armstrong also supplied China with his large muzzle-loaded fortification guns when those Chinese-built guns burst, a blow to Li's hopes for a domestic Chinese armaments industry. Armstrong sent out one of its experts to supervise the manufacture of Armstrong guns in China. Although these were successfully tested in 1879, by then European artillery and naval guns had leaped into steel, breech-loaded technology, thus launching the second wave in the military revolution of the nineteenth century. China was again left behind.[27]

Elswick had to admit they owed Hart a 'special debt of gratitude'.[28] Rendel's role remained crucial in cultivating Hart's friendship and loyalty. He was the company's only stable contact with Hart. Hart succeeded in the 1870s where Lay had failed in the 1860s because he had a different attitude towards China. Whereas Lay had believed that China should put itself in his hands, Hart saw himself and his employees of the Maritime Customs Service as the loyal servants of the Chinese government which paid their salaries. Hart was always concerned that British influence remained paramount in China, but he directed the Maritime Customs Service on behalf of China, not England. His ability and tact gained the respect of the Chinese leaders, with many of whom he developed close and personal relationships.[29] Hart remained Inspector General of the Chinese Maritime Customs Service for over 40 years, turning down – with some hesitation – the offer of the British Cabinet to become the British Ambassador at Peking in 1885.[30] While this certainly helped Armstrong, it was an advantage which would disappear if Hart retired or died. His replacement would have been the German regional commissioner, Gustav Detring, and the position of Krupp would have been strengthened at the expense of Elswick.[31]

In 1880, however, the company sent to China a Major Bridgford, who took charge of the China market, and Hart was used less and less. Stuart Rendel entered Parliament in that year and Hart lost his place in Elswick's marketing system. Bridgford contacted the viceroy, Li Hung-chang, and informed him that armament orders no longer needed to go through Hart. Li and Hart had been engaged in a power struggle over armaments procurement procedures and this was resolved in Li's favour. He now used Armstrong's new agent, Jardine, Matheson and Company. In an awkward meeting in 1880, Bridgford informed Hart of the change.[32] Hart's role as an important Elswick ambassador inside China rapidly diminished.[33]

In 1885, China began to buy modern warships and Hart learned that two cruisers had been ordered from Elswick without his participation.[34] Campbell reminded Elswick that Hart had ordered the first gunboats in 1875 and that the maintenance of a close connection with Hart was still in the interests of the company. The decline of Hart continued, however, for two reasons: the growing power of Li Hung-chang in China, who leaned towards Germany for military armaments, and secondly, the increasing power of Sir Andrew Noble at Elswick. Rendel had developed a close and personal relationship with Hart, while Noble, perhaps because Hart was Rendel's friend, was partly responsible for eliminating him as an Elswick ambassador.[35] For example, when Rendel requested that Elswick send to Hart some technical and price information, Noble did so, but only at Rendel's direct request and without speed or enthusiasm. He sent the material to Rendel, informing him that it was better the information came from Rendel 'as [he was] evidently in high favour with Sir Robert'.[36] But after 1880 Hart was systematically excluded from Elswick's business in China, although Rendel kept him informed of what Li ordered, which only made Hart more despondent.[37]

The self-strengthening movement failed.[38] Naval and military defeats continued after 1861: by Japan in the 1870s, France in the 1880s and again by Japan in 1895 (whose navy included many Elswick warships). China remained a near-helpless object in international power politics and lay at the collective mercy of the world's imperial powers in their scramble for commercial and strategic concessions. The challenges which confronted China in the second half of the nineteenth century proved too much for her. With a powerful, expanding land empire to her north demanding a great army, insistent imperial naval powers on her southern coasts calling for an equally powerful navy, rebellions within and a political structure weakened by rivalries among the provinces, and strong provincial leaders whose loyalties were divided between their local power bases and the central imperial government, China's military and naval situation deteriorated until Japan destroyed it completely in 1894.[39] Li Hung-chang was discredited, self-strengthening never recovered, and the Chinese imperial state collapsed in 1911.

Elswick sold guns to China and attempted to help that beleaguered nation establish an armaments industry because it was good for business. In buying armaments from Elswick, China contributed to the growth of that company and to the proliferation of armaments which underlay the age of imperialism.That it failed to develop its own armaments industry resulted in China becoming a victim of that age. But the failure of China to acquire the armaments with which to defend itself against those imperialist forces with their modern weapons, was due neither to the political ambitions of Robert Hart nor to the political failings of Li Hung-chang, nor to an entrepreneurial failure on the part of Armstrong and Elswick. The roots of China's disaster lay with the Chinese imperial state, which failed to establish close links with firms like Elswick and to accept the political changes which saving China required. In Britain, on the other hand, the age of imperialism brought about a renewal of the military–industrial intimacy of 1854–63.

Notes

1. Joseph Needham, with the collaboration of Ho Ping-Yu, Lu Gwei-Djen and Wang Ling, *Science and Civilisation in China* vol. 5, *Chemistry and Chemical Technology, Part 7: Military Technology: The Gunpowder Epic* (Cambridge: Cambridge University Press, 1986), p. 569; see also pp. 1, 10, 572–9. For the early use of iron and missile warfare in China see Joseph Needham, with the collaboration of Wang Ling and Lu Gwei-Djen, *Science and Civilisation in China*, vol. 4, *Physics and Physical Technology, Part III Civil Engineering and Nautics* (Cambridge: Cambridge University Press, 1971), pp. 682, 687. See also William H. McNeill, *The Pursuit of Power: Technology, Armed Force, and Society Since A.D. 1000* (Chicago: University of Chicago Press, 1982), ch. 2, and Paul Kennedy, *The Rise and Fall of the Great Powers: Economic Change and Military Conflict From 1500 to 2000* (London: Fontana, 1988), pp. 5–10. On the halt of China's voyages of discovery, see E.L. Dreyer, *Early Ming China: A Political History, 1355–1435* (Stanford: Stanford University Press, 1982), pp. 194–205. Chinese technological ingenuity also brought about an agricultural revolution in the fourteenth century: the pedal-driven irrigation pump and the use of fish to fertilise rice paddies drove up the production of rice.
2. Carlo M. Cipolla, *Guns and Sails in the Early Phase of European Expansion, 1400–1700* (London: Collins, 1965); Daniel Headrick, *The*

Tools of Empire: Technology and European Imperialism in the Nineteenth Century (Oxford: Oxford University Press, 1981); William H. McNeill, *The Pursuit of Power*; Paul Kennedy, *The Rise and Fall of the Great Powers*. The same story is found in the history of paper: China invented paper, but it was Europeans who mechanized its production with a vast array of mechanical devices – cranks, flywheels, gears – linked to a water-driven mill. One should keep in mind that Sir William Armstrong was a product of the engineering culture which emerged in medieval Europe.

3. Carlo Cipolla, *Before the Industrial Revolution* (New York: Norton, 1976), p. 207.
4. Kennedy, *The Rise and Fall of the Great Powers*, p. 254 and, for examples, pp. 4–9, 16–30. See also Maurice Bruce, *The Shaping of the Modern World*, vol. I, *Ends and Beginnings: The World to 1914* (London: Hutchinson, 1958), p. 831; Geoffrey Parker, *The Military Revolution: Military Innovation and the Rise of the West, 1500–1800* (Cambridge: Cambridge University Press, 1988); John K. Fairbank, Edwin Reischauer and Albert Craig, *East Asia: Tradition and Transformation* (Boston: Houghton Mifflin Company, 1976), pp. 435–48.
5. The gift metaphor is from Needham et al., *Science and Civilisation*, vol. 5, Part 7, p. 10.
6. Immanual C.Y. Hsu, 'Late Ch'ing Foreign Relations, 1866–1905', in John K. Fairbank and Kwang-Ching Liu (eds), *The Cambridge History of China*, vol. 11, *Late Ch'ing, 1800–1911*, pt 2 (Cambridge: Cambridge University Press, 1980), p. 90.
7. Quoted by Fairbank et al., *East Asia*, p. 561.
8. Another important reformer was Shen Pao-chen (1820–79), Director General of the Foochow Naval Yard from 1868 to 1875, that is, just as self-strengthening was in its early stages. See David Pong, *Shen Pao-chen and China's Modernization in the Nineteenth Century* (Cambridge: Cambridge University Press, 1994). Like Li Hung-chang, Shen saw himself as defending the old Confucian order. The self-strengtheners were not conscious revolutionaries.
9. *The Cambridge History of China*, vol. 11, pp. 61–3.
10. H.B. Morse and H.F. MacNair, *Far Eastern International Relations* (New York: Russell & Russell, 1959), vol. 2, pp. 336–44.
11. Ssu-yü Teng and John K. Fairbank, *China's Response to the West: A Documentary Survey, 1839–1923* (Cambridge, Mass.: Harvard University Press, 1954), p. 69.
12. Wellington K.K. Chan, 'Government, Merchants and Industry to 1911', *Cambridge History of China*, vol. 11, *Late Ch'ing*, pt 2, pp. 422–5; Edward LeFevour, *Western Enterprise in late Ch'ing China: A selective Survey of Jardine, Matheson and Company's Operations 1842–1895* (Cambridge, Mass.: Harvard University Press, 1968), pp. 58–9.
13. Stanley F. Wright, *Hart and Chinese Customs* (Belfast: William Mullan & Son, 1950), pp. 225–6; John L. Rawlinson, *China's Struggle for Naval Development 1839–1895* (Cambridge, Mass.: Harvard University Press, 1967), pp. 34–6.
14. Rawlinson, *China's Struggle*, p. 34; Wright, *Hart*, pp. 225–53. Wright blames the fiasco on Lay's 'ill-advised' attempt to grasp political power (p. 253).

15. Rawlinson, *China's Struggle*; Ting-Yee Kuo and Kwang-Ching Liu, 'Self-Strengthening: The Pursuit of Western Technology', John K. Fairbank (ed.), *The Cambridge History of China* (Cambridge: Cambridge University Press, 1978), vol. 10, pp. 491–542; Jean Chesneaux, *China from the Opium Wars to the 1911 Revolution*, trans. Anne Destenay (New York: Pantheon, 1976), pp. 202–20.
16. Fairbank et al., *East Asia*, pp. 510–11.
17. Fairbank et al., *The I.G.*, vol.1, letter 102, whose footnote identifies B. as Bidwell, an Armstrong agent. The role of Elswick representatives, including Hart, is discussed in the next chapter.
18. Hart to Campbell, 26 June 1874, in John K. Fairbank, Katherine F. Bruner and Elizabeth M. Matheson (eds), *The I.G. in Peking: Letters of Robert Hart*, vol. 1 (Cambridge, Mass.: Harvard University Press, 1975), p. 182 (letter no. 115).
19. T&W, 31/4970.
20. Campbell to Rendel, T&W, 31/4970 (7 August 1879). Campbell wanted Rendel to acknowledge his role in the latter's *Illustrated London News* article (16 Aug. 1879, p. 159) about the latest *Staunch* gunboat built for China.
21. Rendel, *Personal Papers*, p. 248.
22. Campbell to Rendel, T&W, 31/4898 (13 Oct. 1877). Hart's commitment to an Elswick gunboat remained strong. See Rendel, *Letters* (N.L.W.), nos 1741, 1743.
23. For a sample of the 1876 agreements for the recruitment of British officers for the Chinese gunboats, see T&W, 31/4890 and /4891.
24. Fairbank et al., *The I.G.*, vol. 1, letter 126 (11 June 1875). Such a right was always included in the contracts. See an 1880 sample contract, T&W, 31/4993. Hart knew nothing about warship design of course and contented himself with requests for a Gothic rather than an Elizabethan style and reminders that it took three Chinese sailors to do the work of two English crewmen and that five Chinese would fit into quarters designed for four Englishmen (Fairbank, ibid., and letters 140 and 142).
25. Rawlinson, *China's Struggle*, p. 110; Wright, *Hart*, pp. 465–77.
26. Wright, *Hart*, pp. 466–77.
27. Thomas L. Kennedy, 'Li Hung-chang and the Kiangnan Arsenal, 1860–1895', in Samuel C. Chu and Kwang-Ching Liu (eds), *Li Hung-chang and China's Early Modernization* (London: M.E. Sharpe, 1994), pp. 206–8. See also John L. Rawlinson, 'China's Failure to Coordinate Her Modern Fleets in the Late Nineteenth Century', in Albert Feuerwerker, Rhoads Murphy and Mary Wright (eds), *Approaches to Modern Chinese History* (Berkeley: University of California Press, 1967), pp. 105–32.
28. Noble to Rendel, T&W, 31/4892 (3 June 1877).
29. Wright, *Hart*, p. 258.
30. Ibid., pp. 535–42.
31. T&W, 31/3120.
32. Ibid.
33. Wright, *Hart*, p. 477. This was a devastating blow to Hart, whose ambition was to become the central figure in the creation of a modern Chinese coastal navy. Wright (p. 478) contends that Hart nearly resigned and returned to England, but a letter to Campbell indicates otherwise. See John K. Fairbank et al., *The I.G.*, vol.1, letter 139.

34. T&W, 31/2650.
35. Campbell to Hart, T&W, 31/2654 (3 Nov. 1885).
36. Noble to Rendel, T&W, 31/2685 (Jan. 1886).
37. Campbell to Rendel, T&W, 31/3969 (23 July 1894).
38. Emmanuel C.Y. Hsu, *The Rise of Modern China*, 3rd edn (Oxford: Oxford University Press, 1983), p. 285. Hsu lists the reasons for this great national effort ultimately failing: conservative opposition; lack of firm central coordination and direction on one hand and the absence of general participation on the other; the lack of sufficient capital; the fact that the great powers always remained stronger; and the limited vision that industry and weapons were enough, that these 'strengths' could be purchased without a fundamental economic, social and political transformation of China. See pp. 287–91.
39. Rawlinson, *China's Struggle*; Kuo, 'Self-strengthening', pp. 491–542; Kwang-Ching Liu, 'The Military Challenge: the North-West and the Coast', in Fairbank and Kwang-ching (eds), *The Cambridge History of China*, vol. 11, pt 2, pp. 202–73.

PART III

REMAKING BRITISH NAVAL POWER, 1880–1914

Introduction

When Sir William Armstrong approached the Duke of Newcastle with his plans for a new artillery piece in 1854 he was following an ancient tradition of European engineers. Europe's engineers have always been ready to put their skills and knowledge at the service of the state, motivated by a combination of patriotism, a desire to develop their innovative ideas and, by applying engineering principles and skills in new areas, to satisfy their ambition to become known as an innovative engineer. The pursuit of reputation, the desire to accomplish something new and to extend the borders of engineering itself are all powerful motives that have shaped engineering – for good and ill – since engineering began. Archimedes helped defend Syracuse, Galileo taught the art and science of fortification at Padua and Leonardo da Vinci offered the duke of Milan his design for a cannon 'very convenient and easy of transport, with which to hurl small stones in the manner almost of a hail, causing great terror to the enemy from their smoke, and great loss and confusion', along with other technological ideas for 'many engines most suitable either for attack or defence, and ships which can resist the fire of all the heaviest cannon'. Da Vinci designed all sorts of terrible new weapons of unprecedented power: tanks, colossal cross-bows, exploding projectiles, multi-barrelled cannons and machines guns and artillery mounted on a floating platform.[1] But he was unable actually to build these machines because his ideas were ahead of the engineering required, which of course is what drives engineering. Armstrong argued, as engineers still do, that design ideas for an innovation – in his case for light-weight rifled artillery – were plentiful and even solutions to particular technological problems were well known; the difficulty is in the engineering.

Military science and engineering were important and legitimate subjects of study for Renaissance humanists.[2] European chemists have also served their states since gunpowder arrived in Europe in the fourteenth century. Gunpowder armaments prompted the study of ballistics, thus contributing both to the new physics of the seventeenth century and to the new armies of the absolutist states. When the First World War broke out, Fritz Haber, famous for converting nitrogen gas into its nitrogen components and for the vital industrial Haber Process of producing synthetic ammonia, redirected his enthusiasm for finding practical applications for chemistry to weapons. His knowledge of industrial chemistry led him to develop synthetic materials for explosives and poison gas as a weapon.[3] Many German engineers embraced Hitler's new regime because it expanded the opportunities to develop new engineering ideas.[4] The cross-fertilisation between civilian and military science and engineering has been a constant theme in Europe's past.

The direction of engineering and the relationship between civilian and military engineering is complex and shifts according to the changing economic, political and international contexts. The Napoleonic wars prompted Henry Cort, John Wilkinson, Henry Maudslay and Mark Isambard Brunel to work on naval technology, gaining knowledge and skills which no doubt contributed to their later civil engineering accomplishments. After their defeat in 1815, the French Admiralty looked to naval technology to overcome the numerical superiority of the British navy. Paixhan's shell gun and steam-powered ironclad warships were taken up by both revolutionary and reactionary French governments.

Once an armaments technology is developed, transferred and diffused around the world, its impact upon politics increases, and indeed can bend the broader social, economic and political context to its needs. Innovations in armaments can rapidly alter the balance of power between states, and precipitate fears of an invasion or embolden states into pursuing a more ambitious or aggressive foreign policy. The introduction by the French of the steam-driven ironclad battleship in 1859 prompted the British government to shift its home defence from the navy to the army. It was for Armstrong's largest gun that Palmerston built his string of coastal forts. Britain looked to its army and a land-based home defence strategy and allowed the navy to stagnate until 1879, when a war scare with Russia prompted calls for a renewed navy to defend British commerce and colonies. The occupation of Egypt in 1882, the consequent expansion of the French navy and an invasion scare provided the opportunity for Conservative members of the Opposition to attack the government for leaving Britain and her expanding empire undefended. The agitation succeeded in 1884, when William Gladstone, the Liberal Prime Minister who had always opposed increasing armaments spending, was forced to acquiesce in adding £5 million to the naval estimates, beginning the first of several naval arms races that preceded the First World War.

From 1884 until 1907, the naval armaments entrepreneurs enjoyed high reputations as defenders of the nation and grateful governments everywhere conferred honours and titles upon them. Armstrong was made a peer of the realm and given a seat in the House of Lords. In the 1890s, the new industrial navies became very popular, and the navy celebrated with jubilees, reviews, coastal cruises and exhibitions. A Navy League and an Imperial Defence League were founded to organise and sustain popular agitation on behalf of the navy. The naval histories of

Alfred Mahan, an American navy officer, helped convince both professionals and the general public that command of the sea was crucial to defence of home and empire.[5]

But as the cost of battleships rose and naval arms races accelerated, the role of private armaments companies in promoting armaments came under question. The growth of naval armaments appeared to be less about national or imperial defence and more about the financial interests of private iron and steel companies.[6] The pivotal year was 1907. Until then, for example, *The Economist* accepted rising British naval estimates as an appropriate response to German foreign policies. After 1907, the magazine gave Germany the benefit of the doubt, and attacked British naval programmes as financially and strategically unwarranted.[7] The great divide came in 1909, when the claims made by the managing director of the Coventry Ordnance Company that Germany was secretly accelerating its naval building, gave the Conservative Opposition another opportunity to accuse the Liberal Government of ignoring British naval defence. As in 1884, this agitation succeeded too, and the government increased the naval estimate to build eight battleships rather than the four originally planned. When the claims about the German programme were proved false, private armaments companies were accused of spreading false information to foment war scares.

The Balkan Wars of 1912–13 were seen as evidence of 'a great and growing evil' link between armaments companies, banks and government diplomacy of the great powers, which forced loans upon bankrupt states with which they were obliged to buy armaments.[8] Karl Leibnecht, a Marxist member of the German Social Democratic Party, spoke in the Reichstag of bribery and corrupt tactics used by Krupp, and denounced 'the bloody international of the merchants of death'.[9] He claimed Krupp had bribed German officers to obtain War Office files.[10] He read out a letter from a German machine gun manufacturer instructing its Paris agent to have an article published in a widely-read French newspaper claiming that France was accelerating production of machine guns.[11] Just as Mahan's celebration of naval power had raised the popularity of navies and the private companies that built great battleships, Leibnecht's famous revelations sparked an outburst of attacks upon naval arms races and the private firms who were seen to be behind them. The scandals in Germany prompted British journalists and radicals to investigate their own arms companies. In 1914, a great scandal broke out over the bribing of Japanese officials by Vickers. Armament salesmen often played traditional enemies off against each other as a means to increase orders: Chile against Peru to sell them both submarines; Brazil against Argentina to sell them both battleships. Similar tactics of bribery and fear were used in Turkey.[12]

These kinds of sales tactics demonstrated to many that the commercialization of armaments meant that the maintenance of production capacity through international sales of armaments superseded diplomatic agreements or even strategic calculations. The commercial production of armaments put profits ahead of patriotism, and thus private armament companies sold to any state, including potential enemies. The critical attacks on arms firms sharpened with each new revelation of price fixing, arms trusts and international rings which divided markets and shared technologies.

When the First World War proved to be longer and less glorious than expected, and one in which battleships played an insignificant role, the cause of the war became an increasingly important question. A variety of conflicting reasons were put forward which blamed European governments, but one cause achieved a consensus: that private manufacturers, in their relentless search to expand markets and profits, created the emotional conditions which led to war. In his postwar memoirs, Viscount Grey, the British Foreign Secretary from 1905 to 1916, concluded that 'the moral is clear, great armaments lead inevitably to war'. The expansion of armaments in the years leading up to 1914, he said, did not strengthen nations, provide security and promote peace, but rather generated fear, suspicion and distrust among nations, and the more destructive the armaments, the greater the fear and distrust, all of which promoted 'evil imaginings of all sorts' which poisoned diplomatic negotiations. The diplomatic failures of 1914, he said, did not provide 'the real and final account' of the origin of the First World War. 'The enormous growth of armaments in Europe, the sense of insecurity and fear caused by them – it was these,' said the former Foreign Secretary who had played a leading role in that growth, 'that made war inevitable.'[13]

The persistent strength of the belief in the link between arms races and war is demonstrated by the Washington Treaties of 1921–2, by which the major powers, in an unprecedented move, voluntarily limited their naval armaments.[14] At the same time, the League of Nations launched an inquiry into the question of the ways in which wars were 'promoted by the competitive zeal of private armament firms and would be rendered less frequent were the profit-making impulse brought under control or eliminated altogether'. Individual armament firms, it was claimed, fomented war scares and arms races by disseminating false, exaggerated or distorted reports about the military–naval preparations of rival nations.[15]

The armaments question faded after the Washington Treaties of 1921, the 'peace of Locarno' of 1926 and the Kellogg–Briand pacts of 1928, but re-emerged as a topic of intense public debate when Hitler withdrew from the Disarmament Conference and the League of Nations in 1933, and launched his rearmament programme in 1935. The role of armaments in raising the probability of war was an ideological pillar of appeasement, the foreign policy of Britain at the time, and the previous accusations against armaments companies found a wide audience. Private armaments companies were once again accused of fomenting war scares and playing off nations against each other and virtually forcing governments to buy unnecessary quantities of weapons. Sharing international markets, setting prices and production quotas, and selling weapons to all sides at war outraged (and adumbrated decisions made in 1914) the patriotism of the main critics of the armaments business, radical liberals and socialists. The new international armaments business was an affront to traditional notions of national capitalism which was presumably restrained in its quest for profits by competition and patriotism. This was the core of the crime, and it took on greater proportions when the management and shareholder lists of these companies revealed the names of admirals, generals and politicians. These men – and the few women who held shares in arms companies – provided the social and financial foundation of the 'merchants of death'.[16]

Proponents of the merchants of death thesis made some telling points about the political influence of armaments industrialists. There were national and international agreements amongst arms companies of which governments were aware, and there were also a few conspiracies by arms salesmen to promote war scares or naval panics amongst their customers. But the cause of the First World War cannot be boiled down to commercial armaments. Nevertheless, neither the Armstrong, Whitworth Company nor Vickers could survive the post-1918 context. No entrepreneurial skill could save them, and by 1926 both were bankrupt.

The Second World War ended the debate, but the Cold War context of the 1960s brought the question of private armaments and war to the surface once again. Those who discounted the merchants of death thesis accepted that such tactics as bribery and war scares were in fact practised by private armaments companies.[17] But they rightly emphasised that the merchants of death as a thesis has limited scope to analyse the international commercial armaments market in general or the armaments industry as a whole. However interesting or true the scandal stories, bribery and scare-mongering cannot account for the growth of a global armaments business. To be sure, as more companies entered the field, and as states developed their own arsenals, competition in the global armaments market became fierce, but to portray armaments companies as merchants of death is to wildly exaggerate the overall importance of bribery and tricks, which tended to cancel each other out, leaving no firm with an overall advantage. There is no evidence that Armstrong participated in any corrupt behaviour, and no evidence that the firm suffered because it refused to employ such marketing tactics.

A second point about the merchants of death theme involves its focus on the few years just before the First World War. Virtually nothing is said about the marketing techniques of private armaments companies in the period before 1900, well before they became merchants of death, when they were still entrepreneurs and defence industrialists. Commercial armaments did play an important role in bringing about the First World War, but not in the simple way suggested by the merchants of death thesis as it stands. The selling of commercial armaments in a global market in the 1907–14 period was shaped by the decisions of states, and was part of the international diplomatic game.[18] Clive Trebilcock described what was called the 'special relationship' between governments and arms companies, one resembling a partnership in which the government was the senior partner and whose power as both manufacturer and buyer outweighed the power of arms companies as suppliers. British governments, he said, remained firmly in control of both the design of new weapons and the distribution of orders among the private firms and the prices that were charged. Governments, through their control of production quotas and prices, he argued, established and maintained healthy competition amongst the arms firms.[19] Robin Higham concluded that the most salient feature of the relationship was 'the omnipresent nature of the government as customer, paymaster and regulator' which exercised 'virtually monopolistic powers, forcing companies to tailor themselves to its demands and the whims of its servants'.[20]

The relationship between private armaments companies and their own governments was complex, but historians generally fall into two general camps, one represented by James Joll who, with his focus on the few years just before 1914, concluded that 'in general ... governments needed arms firms more than the arms firms needed governments',[21] and the other argued by Arthur Marder who, looking at imperialism in the 1890s, concluded that 'munitions makers were only supplying a demand ... they were in themselves only the instruments of policy'.[22] Thus, as with many historical problems, there are as many answers to the nature of the relationship between private armaments companies and governments as there are contexts and periods of time from which the problem can be perceived. Both Joll and Marder could be correct because the balance of power in the private armaments industry, and especially between specific firms within the naval armaments sector, fluctuated according to the state of technological development achieved by the private firms, the size of foreign demand and the intensity of the international rivalry.

All of these contextual factors were altering dramatically and rapidly after 1880. At the end of the nineteenth century, the traditional focus on the balance of power in Europe was challenged by the new forces of imperialism and *Weltpolitik* which had been made possible by railways, navies and modern artillery. As the world diplomatic crises multiplied, the structure of the arms trade changed, becoming more and more dependent on the domestic demand, and less on foreign sales. British naval armaments companies entered a period of rapid growth when Britain began to increase and modernise its navy in 1885, followed by a burst of growth and new entrants with the great Naval Defence Act of 1889. But rapid growth and rising profits came at the cost of entrepreneurial control and independence by the private manufacturers to the state. The customer was close at hand and the orders large, but it was a market that depended too much on a single buyer. As private arms companies put themselves under greater government control, the rewards were high, but the long-term dangers were great. Armstrong was brought down by this in 1863, but then he could turn to the rising global market. When British orders stopped after the First World War, the Washington Treaty of 1922 left no new markets to which the company could turn.

The broad political and international contexts became an increasingly important part of the armaments business after 1880. The international–diplomatic context of a growing rivalry amongst the great powers clearly pushed up the demand for warships, and prompted a great wave of design innovations in warships and weapons. A second consequence of the rise in naval expenditures was a tightening of the relationship between the British state and the private manufacturers of armaments. The control of the British government over private armaments production increased as it became overwhelmingly the single most important customer. The government followed tradition in calling upon the private trade when international emergencies arose (in this case an increase in the number of French and Russian warships between 1880 and 1884, and the formation of the Franco-Russian Alliance between 1888 and 1894).

Britain responded to these naval threats with major naval building programmes in 1884, 1889 and 1894. Once again the government turned to private iron and steel companies to produce what ships and guns its own arsenals and dockyards could not. The government encouraged iron and steel manufacturers such as Vickers in 1888 and Coventry in 1906 to move into armaments production with promises of naval orders. To keep prices low the government also used its monopoly of demand to foster competition among the armament companies. At the same time, however, it discouraged technological innovation in guns and warships by imposing its own designs on the private firms, making them mere assembly plants. Consequently, in the 1890s, Armstrong faced a fundamentally new structural context in which demand narrowed from one predominated by a number of global markets to one of a virtual monopoly by a single state; competition was increased by the government and yet technological innovation was discouraged.

The relationship between private armaments firms and the British state thus rested on a tension between the desire of the state for naval armaments and the need of private manufacturers for profits. The British state became a sort of general manager of the armaments industry, providing diplomatic support for foreign sales, determining demand, creating competition, setting technological designs and prices. In short, a number of entrepreneurial decisions shifted from the company to the state. Between 1884 and 1914, there arose an intimacy between Armstrong's company and the British state, a complex relationship of mutual dependency, acrimony and cooperation. It was a period of contradictions: the company lost much of its entrepreneurial independence in exchange for tremendous growth in the service of the state and a place at the table with prime ministers and foreign secretaries. After 1890, the capacity to build warships quickly replaced technological innovation as the main factor in success in the armaments business, yet strategic and diplomatic discourse and decisions became increasingly technological. Although marked by conflict, the interests of armament companies had become the interests of the state, and diplomacy reflected that change. The border between national interests and business profits blurred, the distinction between foreign policy and strategy on the one hand and the economics of the armaments industry on the other melted away.

British governments sponsored the industrialisation of naval armaments, passed increasingly large estimates to build more warships, ended the state monopoly of weapons production and provided diplomatic support to private companies competing for foreign sales, and created a naval–industrial complex, all in order to uphold their traditional fundamental foreign policy goal of maintaining diplomatic freedom of action in the balance-of-power system of international relations.

International rivalries, armaments technology and domestic political struggles were reinforcing processes in the arms races between Britain and France from 1840 to 1863, Britain's naval programmes against France and Russia from 1884 to 1900, and the great race for dreadnoughts between Britain and Germany from 1906 to 1914. Each of these processes affected the others, but each also possessed its own, relatively autonomous, momentum. Certainly, international relations based on the balance of power encouraged technological innovations in armaments, but particular armaments were the product of creative engineers struggling with the limits which available materials imposed on technological design. The external context cannot simply call up any technological innovation it thinks necessary, but neither do technological innovations move only according to some inner technological logic of its own. It is the evolving and changing relationship between technology, business and politics that is at the core of modern world affairs and why the history of Sir William Armstrong is an important historical agent.

The story of Armstrong and his company is thus part of a larger story of the decision of governments to link their diplomacy more closely to fast-changing military and naval technology, a decision which over time weakened the ability of diplomacy to control events. On 18 March 1914, Sir Edward Grey gave his opinion of the old adage that armaments depended upon policy. He claimed that the statement 'policy must have some relation to armaments', was at least as true, and hence British foreign policy must be developed with naval strength in mind.[23] The nature of the relationship between foreign policy and the military or naval power available to enforce it had been a major point of debate in British politics. In 1901, Campbell-Bannerman had struggled to retain his belief that, in the end, 'policy governs armaments', but recognised that 'the policy and the armaments act and react upon each other', and an increase in armaments could sway policy.[24] The following chapters attempt to chart the ebb and flow of the ambiguous relationship between Elswick and the British state during the critical years of 1880–1914.

CHAPTER SEVEN

The naval–industrial complex

I

The knowledge and techniques developed during the first phase of the Industrial Revolution, running from about 1760 to 1840, were applied to heavy gun and warship production by men like Armstrong, Krupp and Whitworth between 1850 and 1880, and laid the foundation for the development of modern industrial armaments companies. As we have seen in Part I, guns and warships went through a period of rapid technological development in the 1860s and 1870s, yet for all the dramatic expansion of these two decades, during the period 1880 to 1914 even greater and more rapid technological innovations took place in the armaments industry. This new generation of weapons was part of the second phase of the Industrial Revolution, which began about 1880, and during which chemistry, electricity, metallurgy and machine engineering made enormous strides far too sophisticated and extensive to be discussed here at the level of detail that was possible for the period examined in Part I. However, the general point remains that weapons technology evolved mostly from the research and development efforts of private entrepreneurs. Governments contributed relatively little to the major innovations. They encouraged their arsenals to make cheaper versions of those innovations which could not be ignored and were content to play the role of customer to entrepreneurs and took no interest in the plight of private manufacturers whose production costs grew while government orders fluctuated according to the oscillations of international relations.

In the 1880s, the costs of developing and manufacturing modern naval armaments escalated, and the aloof attitude towards the private armaments industry held by both Liberal and Conservative governments since 1863 could no longer be sustained against the political pressures generated by a new international order and the revolutionary innovations in armaments technology that poured forth after 1878, led by cheap steel and slow-burning gunpowder and advances in armaments engineering. At the same time, while Elswick continued to make profits in the overseas markets, the economic depression that began in 1873, along with the relative decline in the number and scale of armed conflicts, made future growth uncertain for the increasingly costly guns and ships. Since 1863, Armstrong had looked forward to renewing Elswick's association with the British government.

The strategic decisions confronting the company centred on the kinds of armaments and warships to manufacture the transformation of the firm into a public company, the ways and means of managing growth and expansion (vertical integration and/or horizontal alliances or amalgamations, issues of corporate governance, and the development of a close relationship with the government). The technological decisions were the least controversial. The fundamental principle remained Armstrong's belief that the gun would always prove superior to armour and therefore warships should be designed to accommodate the latest developments in ordnance. To Armstrong a warship, however complex, was but a floating gun carriage, and thus the design of warships must always be determined by the evolution of the gun. In the 1860s and 1870s, this meant ships were designed to carry the ever-increasing weight of the latest monster guns. The weight of armour on a ship must, he argued, be subordinate to the requirements of the gun. In 1867, this principle was put into practice with the launching of the first *Staunch* gunboats, designed by George Rendel and constructed in the shipyard of Charles Mitchell. Armstrong sold these small, unarmoured vessels with their powerful 35-ton rifled muzzle-loaded gun as an effective answer to the great lumbering ironclads whose prices were rising sharply. The *Staunch* carried a gun of equal size to those on board the largest battleship but was a small and fast-moving target. Its low draught allowed it to enter shallow coastal waters or to go up rivers to escape a battleship. Finally, as Armstrong continually pointed out, he could build a dozen of these small but formidable ships for the price of one ironclad.[25]

The *Staunch* was a tremendous success for Elswick. As we have seen, its low cost made it attractive to customers like China who could not afford to buy ironclads but needed some degree of naval defence. The Chinese navalists no doubt felt a renewed sense of security from the observation of one British commentator on their purchase of the latest *Staunch* who said that China, had 'by a sudden and adventurous leap placed herself abreast of ourselves'. The gun of the *Epsilon* was viewed in Britain as 'the most formidable weapon which the English navy possess at the present moment'.[26] By 1877, Elswick was capable of shipping dismantled *Staunch* gunboats in crates and reassembling them at their destination under the supervision of Elswick's experienced technicians.[27]

The British Navy bought a few *Staunch* gunboats in 1867. The Under-Secretary of War (Lord Longford), the Attorney-General (Sir Hugh Cairns) and Henry Corry, First Lord of the Admiralty, were among

Armstrong's close political friends.[28] Cairns had been a fellow-student of Armstrong's at Lincoln's Inn in the early 1830s and in 1858 had helped bring in the amendment to the patent act which had enabled the government to keep Armstrong's gun a secret. He had also been instrumental in countering Blakely's claims against Armstrong.[29] Corry too had taken up Armstrong's interests in Parliament[30] and he ordered the first *Staunch* gunboats for the British Navy. The Admiralty, however, continued to emphasise British high seas naval power and, casting off their previous conservative attitude towards technological innovation, pressed for money with which to build ironclads. But their naval scare tactic of 1868 made as little impression on Disraeli as earlier efforts had made on Gladstone.[31] As one historian remarked, Richard Cobden exercised more influence on British foreign and naval policy from his grave than he had ever done from his seat in the House of Commons.[32] This would change in the 1880s.

British governments of both parties remained loath to spend money on new armaments, and research and development of new guns was left to Elswick. Cabinet retrenchment was complemented by the conclusion of the army that even small breech-loaders remained technologically inferior to muzzle-loaders. The 1870 decision by the British army to abandon breech-loaded artillery and return to muzzle-loaders was based on technological and financial considerations which reflected the budget-concious government of William Gladstone.[33] See pp. 87–93 on the army's rejection of Armstrong's breech-loaders.

Although Armstrong continued to experiment with breech-loaders, the company remained committed to the development of the largest and most powerful naval gun, and that meant muzzle-loaders.[34] Large breech-loader ordnance had to await further advances in gunpowder and it was to this that Elswick devoted its research and development energies during the 1870s. Two men worked on it: Sir Frederick Abel and Andrew Noble.

Abel had studied under the Austrian chemist, August Hofmann and, following a variety of lectureships in England, was appointed Ordnance Chemist at Woolwich in October 1854. England and France had entered the Crimean War in March and Abel designed underwater mines to blow up the sunken Russian ships which clogged the entrance to Sebastopol Harbour. Armstrong had gained a reputation in the early 1840s with his experiments in hydro-electric systems and he was commissioned by the War Office to design an electrical detonation device for the mines. After the war, Abel built an extremely successful career upon his knowledge of explosive powders and held numerous positions: War Office chemist; professor of chemistry; president of various governmental committees on explosives; president of various scientific institutions (he was chairman of the 1877 meeting which founded the Chemical Society, and its first president); president of the British Association; and many more.[35]

Between 1854 and 1893, Abel was the War Office chemist and its chief advisor on explosives. At the War Office, he had access to new ideas sent in by private inventors.[36] In 1865, Abel himself attempted to negotiate a contract to supply the War Office with guncotton, but received 'scanty encouragement' from the War Office: he worked for the War Office and a new rule prohibited government employees from holding patent rights on items bought by the government. Abel had to resign and attempt to sell the guncotton to the War Office as a private manufacturer, but War Office authorities held 'strong prejudices' against buying from private manufacturers. He asked Armstrong for advice and his old friend suggested he either give the patent to a third party or sell it to the government for a lump sum payment, as others had done.[37]

Abel remained closely associated with Elswick and became one of the first scientific consultants to the modern armaments industry.[38] Abel and Noble began their experiments in 1877 in the laboratories of Elswick and developed a new slow-burning powder which laid the foundation for the second wave of the revolution in ordnance. The old powder exploded instantly but partially, and the breech mechanism could be destroyed after only a few shots. The new powder allowed a controlled and thorough burn to occur, thus imparting a continuous force to the projectile throughout the length of the barrel. Longer burns produced dramatic increases in muzzle velocity, which jumped from 1600 to 2100 feet per second. Longer barrels were developed as the limits of this new system were sought.[39] The work was carried out in Elswick laboratories and the patent rights stayed with the company. Abel did not join the company made possible in part by his research but, 'on the special advice of [his] old friend Captain Noble', he bought 150 shares in the new company which this development encouraged Armstrong to form.[40]

Meanwhile, the age of steel was under way. Steel and the new explosives revolutionised the armaments industry by making possible large breech-loaded guns and armour that was far tougher than iron. Armstrong had tried a steel tube as a liner in his first gun in 1855, but at that time the homogeneity of the metal was uncertain and Armstrong decided on an all wrought-iron gun.[41] Bessemer, attracted to armaments technology by the destruction of the Turkish fleet at Sinope and the Crimean War, set about developing a steel shell and gun metal. In 1859, he convinced Colonel Eardley Wilmot, the superintendent of the Royal Gun Factory, to install a converter at Woolwich, but within a week of the initial agreement Armstrong replaced Wilmot and Woolwich was given over to wrought-iron guns. Bessemer became another victim of Armstrong.[42]

Steel is distinguished from pig iron and wrought iron by its carbon content. Carbon determines the malleability, hardness and strength or durability of iron metal; the greater the amount of carbon, the harder (but more brittle) the metal. Pig iron contains between 2·5 and 4 per cent carbon and is very hard but brittle. It breaks if worked or put under stress. It is thus cast into moulds to produce heavy but usable products such as pots, pans, fences, railway tracks or muzzle-loaded, smooth-bore guns. Wrought iron, with a carbon content of less than 0·1 per cent, is soft and can be shaped into building materials or, as Armstrong first demonstrated in 1855, worked into breech-loaded artillery pieces. However, the price of high malleability, as Armstrong discovered with his 110-pounder naval gun, is short-lived durability and low resistance to shock. Steel has carbon content between that of cast iron and wrought iron and is thus hard but elastic, resistant to percussion and has strength in proportion to its weight and volume far greater than that of iron. Steel is light, hard and malleable, and was the best available material in 1880 for hammers and anvils, rails and locomotive wheels, skyscrapers, ships, armour and guns.[43]

Reliable steel at prices competitive to iron was developed over the 20 years following Henry Bessemer's famous converter process, announced by him in 1856, which blasted air through the molten iron and burned off the carbon as carbon dioxide in about 20 minutes, a method which, while expensive, was far more efficient than puddling. However, the Bessemer method did not eliminate phosphorous, another ingredient in many iron ores which rendered steel useless for many industrial purposes, especially armaments manufacturing.

William Siemens, a German engineer living in England and working on ways to retain heat, developed the open hearth method by which the liquid iron was placed in a shallow pan and subjected to a stream of hot air and a flame of gas, a process he used first for glass making and then applied to steel production. The Siemens method, a rival to the Bessemer process, was discussed at the 1868 meeting of the British Association, where both Bessemer and Armstrong praised it.[44] However, the phosphorous problem persisted and ironmasters remained reluctant to make large investments in steel production. Many worked on the problem and it was solved in 1878 by Sidney Thomas and his cousin, Percy Gilchrist, who eliminated the phosphorous from a batch of steel made by the Bessemer process. Thomas and Gilchrist are rightly given credit for finally launching the steel age, but the Siemens–Martin open hearth method of making steel was more responsible for introducing steel into armaments manufacturing. With the help of two Frenchmen, Pierre and Emil Martin, William Siemens developed the open hearth process which produced steel that was virtually free of phosphorous. The procedure required 14 hours, compared to the Bessemer steel which took only 20 minutes to produce. However, for armaments manufacturing, this was an advantage since the longer period of time provided the opportunity to test and adjust the chemical content of the molten metal during the refinement process and ensure that the steel had the precise characteristics critical in armaments.[45] In the 1880s, steel alloys of nickel, manganese, silicon, tungsten and chrome were developed which made armour plate tougher with less weight and opened another round in the competition between guns and armour.[46] Thus Armstrong chose the Siemens–Martin process to make the steel for his new generation of guns and ships.

Production of the new steel required a very large capital investment and, in 1882, Armstrong carried out his first major amalgamation and, at the same time, restructured the old partnership into a limited liability company. Armstrong bought the Low Walker shipyard, owned by Charles Mitchell and Henry Swan, where the *Staunch* gunboats were built, and Mitchell and Swan joined the board of directors of the new public company, the Sir W.G. Armstrong, Mitchell and Company.

The Low Walker yard made civilian ships as well as Armstrong's gunboats, hence the merger also put Armstrong into the merchant ship building business. But there is no doubt that the new company was created to extend warship production. Although the expansion was to take place 'without encroaching upon the building of ships for the mercantile marine and for passenger traffic', the *Prospectus* made it clear that the civilian ship building department of the company would receive little executive attention: 'great and increasing as is the demand for steel for shipbuilding purposes, it is especially in relation to gun-making that steel works are required'. The *Prospectus* emphasised that the future of Sir W.G. Armstrong, Mitchell & Co. rested on the 'new era in artillery [which was] marked by the recent adaptation of guns for enormously increased charges of powder, an adaptation which was initiated by the Elswick Company in the year 1877'. Along with the recent removal of 'the uncertainties and imperfections of steel', the new powder made possible the construction of ordnance of unprecedented power. The *Prospectus* modestly boasted that 'the superiority of modern guns is so great that extensive re-armaments may be anticipated both in this country and in others'.[47]

The nominal capital of the new company was declared to be £2 million in the form of 20000 shares of £100 each. Assets were valued at £1575000.[48] The vendors of the two merging companies (Armstrong, Mitchell and their respective partners) took £1185000 in 11850 fully-paid shares and £390000 cash, structured as a loan at 5 per cent interest to be paid out of future profits in four equal payments. The company retained 1500 shares and the remaining 6650 were offered to the public. Investors could buy a share in five instalments between November 1882 and June 1883. Those who immediately paid in full received 4 per cent interest during that period. Based on the performance of the two companies over the previous five years, the company estimated a return on investment 'of fully 10 per cent', and 'nearly as favourable' if the previous ten years were considered.[49] The public offering thus raised £665 000 to finance the new steel plant and shipyard at Elswick.

A new Elswick warship, a cruiser designed upon the *Staunch* class gunboat, was introduced in 1882. The *Esmeralda* had two 10-inch breech-loading guns and six 6-inch guns. With less than 4000 tons displacement, it attained over 18 knots speed. There was no armour and the engines, boilers and magazines were inside steel decks below the waterline. Armstrong claimed the ship was 'almost absolutely secure against the worst effects of projectiles'. It was, he said, 'the swiftest and most powerfully-armed cruiser in the world' and, in a comment directed towards the British Admiralty, added that 'no cruiser in the British navy is swift enough to catch her or strong enough to take her'. The first *Esmeralda* had been bought by Chile, a nation unlikely to be at war with England. And a good thing for Britain, said Armstrong, because he could 'conceive no more terrible scourge on our commerce than she would be in the hands of an enemy'.[50] As with the sales of the monster guns to Italy and the *Staunch* gunboats to China in the 1870s, the sale to Chile of the *Esmeralda* cruiser offered Armstrong the opportunity to warn that, without Admiralty orders, he might be forced one day to sell to potential enemies.

Armstrong's forecast of demand for the new guns and ships was correct. The Engine Works had put on a night shift to help the Ordnance Department as foreigners were impatiently waiting to buy the new warships.[51] In October 1884, the new shipyard at Elswick was finished and *Esmeralda* cruisers began pouring out: two for Austria, one for Italy, others for China, Spain, Romania, Argentina, Norway, Portugal, Turkey, Brazil and one to the United States. A wave of new technologies swept across the industry. By 1889, triple expansion and turbine engines raised the speed limits and in 1894, Armstrong built for Japan the battleship *Yashima* with a speed of 20 knots despite its 12 000-ton displacement and enormous 12-inch guns.[52]

World demand for *Esmeralda* warships was fuelled by their cheapness. At less than £400 000 they were under half the cost of a contemporary battleship. This dynamic market for cruisers lasted throughout the 1880s, fuelled by the belief among many naval strategists that the days of the battleship and 'command of the sea' were over and the era of submarines, fast cruisers and wars on commerce had begun.

Figure 7.1 The Armstrong cruiser Esmeralda, *National Maritime Museum, Greenwich*

Esmeralda cruisers were as fast as torpedo boats and carried quick-firing guns to destroy them. It was a decade in which the kinds of ships Elswick produced were in highest demand, a time when Armstrong's ideal combination of heavy guns on fast unarmoured cruisers reached a peak in their historical development.

II

Armaments were a major part of the technology and industrial developments associated with the truly modern phase of the Industrial Revolution that had begun a hundred years earlier. It is difficult to exaggerate the impact cheap but high-quality steel and high explosives had on armaments and on Elswick in particular. Armstrong moved into steel guns immediately and his firm made contributions to the development of high explosives. There was no lack of entrepreneurial spirit when it came to pursuing technological innovations in Elswick's laboratories, and no hesitation to expand production facilities to take full advantage of those technological breakthroughs.

By 1880, the structure of the arms business had changed dramatically too, much of which was a consequence of Elswick's technological innovations and the financial power which came with dominating the international markets in the 1860s and 1870s. In 1854, dozens of private companies could build the big cast-iron, muzzle-loaded smooth-bored guns within just a few months of the British government ordering them in the sudden crisis in the Crimea. This was no longer true in 1880. The shift to the huge wrought-iron, rifled guns which Armstrong manufactured wiped out the vast majority of those firms by making their large and expensive foundries obsolete, and by driving up the costs of armaments production, which now required special and expensive machines to rifle guns, make the coils and the increasing number of components that heavy guns now required. The shift from design to production involved very large investments and the engineering knowledge and financial capacity. The design concepts were well known; it was the engineering that posed the real difficulties. Since 1854, Armstrong's success came from his ability to overcome the engineering problems of manufacturing modern armaments. Production engineering became increasingly complex as the size of guns increased quite markedly during the 1860s and 1870s with the new introduction of steel breech-loaders and the slow-burning powder developed by Noble and Able in Elswick laboratories.

By 1880, the government arsenal at Woolwich was far behind in design, engineering and production. It continued to make wrought-iron smooth bores, and did not have the capacity either to design or to manufacture the new steel guns. The old guns could not withstand the strain of the slow-burning powder and elongated steel projectiles. Private firms seeking contracts from Woolwich were discouraged from submitting innovative designs because Woolwich was unable to make them, and they would not sacrifice their engineering monopoly to design changes. Nor would they consider the French method by which armaments were made by private firms under supervision of government inspectors. Armstrong could not sell to the War Office or Admiralty because he did not make what Woolwich made, and Woolwich would not buy anything from private manufacturers unless they could make it also.

This old scheme of 1859–63 was completely outdated by 1880. When the next international and political crisis came along, the old method of calling upon private firms to fill the orders that Woolwich was unable to handle would not be possible because what Woolwich made was of inferior materials and outdated design, which the private firms did not make any more. Moreover, the new guns could no longer be manufactured within a few months. Armaments production now required time and careful planning and coordination with ship building. But the realities of naval armaments could only be ignored for so long. In 1885, the government was forced to turn to Armstrong for design and production engineering for naval guns. The balance of power between Elswick and Woolwich shifted to Elswick in 1885, when Sir William Armstrong was asked to join the new Committee on Designs for recommendation of the design and production of the new steel rifled breech-loaded naval guns. The relationship with the government is taken up in the next chapter, but here one may add that this acceptance by the government of the private armaments industry which began with Armstrong was not made the foundation of Britain's defence industry, as Stuart Rendel so strongly advocated. British governments were always ready to sacrifice the private firms to maintain Woolwich, and this was done when the great crisis that began in 1885 ended in 1918, to the great detriment of British national defence.

By 1885, Elswick's new breech-loaded steel guns and their hydraulic machinery for firing, its new quick-firing guns, its new fast heavily-armed cruisers, could no longer be ignored by the British Admiralty, and demands increased inside the Admiralty to overturn the Woolwich monopoly and allow the navy to purchase weapons from the private sector, that is, from Elswick. The second generation of armaments developed by Armstrong destroyed any remaining pretensions that Woolwich was an innovative arsenal. The technological sophistication of modern weapons rendered impracticable the old method of stealing from independent inventors. The War Office continued to refuse to supply any new weapon that it did not make itself, but the new weapons brought into focus the technological backwardness of Woolwich.

The international context, especially the emergence of the French strategy of war against British commerce – the *Jeune Ecole* school[53] – brought this pressure to a climax in 1884–5. The Admiralty, in response to this threat, demanded the right to buy guns directly from Elswick.[54] The first successful assault on the War Office's monopoly was carried out by Admiral Fisher in 1886, when he gained the right to purchase the large-calibre quick-firing gun for use against torpedo-boats, developed by Joseph Vavasseur at Elswick in 1881.

The Royal Gun Factory had fought against the intrusion of Elswick for two decades. In 1863, it manipulated its accounts to make it appear that Armstrong guns cost less to manufacture at Woolwich than at Elswick.[55] In 1867 and 1868, a similar conflict erupted over the costs of converting breech-loaders into muzzle-loaders by a method developed by Sir William Palliser, who had arranged a royalty agreement through Rendel to have Elswick carry out the conversions and the manufacture of Palliser projectiles.[56] This arrangement was destroyed by the accounting machinations of the War Office.[57] The War Office sacrificed quality and innovation to price: a cheaper version of the Armstrong gun was manufactured at Woolwich from a design of a Captain Fraser. After 1880, however, Elswick had pulled so far ahead of the government factory that the old pricing manipulations worked against Woolwich and increased the profits of Elswick. At the hearings of the 1887 Committee on Army and Navy Estimates, Lord Randolph Churchill successfully challenged the War Office claim that it manufactured large guns more cheaply than the private trade.[58] After 1880, Woolwich fought a long but losing struggle against Elswick over the system of armaments production in Britain.

III

The first battle between Elswick and the War Office in the 1880s was over the proposed construction of a new steel plant at Woolwich. Elswick was in the midst of investing heavily in its own steel mill and Stuart Rendel launched a vigorous campaign against Woolwich. In 1881, he attacked the reactionary and monopolistic attitudes of the War Office which had been the major impediment to the technological advancement of English guns. Woolwich, he said, was not interested in new and better guns, but only in maintaining its monopoly over the supply of guns for the British army and navy. It played off rival civilian inventors against each other and then eliminated the winner. In the 1860s, 'Whitworth was got rid of by playing Armstrong off against him; Armstrong was ousted by shifting upon him the exclusive responsibility for all its failures.'[59] The stakes – the large capital investment, the expanded size of Elswick, its new role as a limited liability company – were now much higher than in 1863, and Rendel was determined that the experience of 1862–3 would not be repeated. The refusal of Woolwich to encourage innovations from Elswick had helped Krupp gain in international reputation at the Meppen demonstrations in 1879, with technical developments which were 'mere plagiarisms upon Armstrong's models'. The 'deadening influence of Woolwich,' said Rendel, was destructive to both England and Elswick, whose interests were the same in his view:

> How could a private English firm hope to secure a substantial business with foreigners in guns which the English Government deliberately neglected? Thus Krupp, as the rival less obnoxious to Woolwich, was suffered to carry off both the credit and profit of the new step in ordnance.[60]

The new steel, breech-loaded ordnance technology and the emergence of Krupp put the future of Elswick in question if Woolwich retained the power to fix designs, control all British orders and systematically ignore any new technology which it could not steal. In a remark which had been made by many inventors in the 1850s, Rendel accused the Ordnance Committee of maintaining the 'prestige and power of the Government manufacturing establishment' by forcing all private inventors to reveal their secrets to the committee. Patent laws allowed the War Office to confiscate virtually any knowledge or technique submitted to it by private inventors. Elswick had avoided this by selling in the world market, but now that was threatened by Krupp who enjoyed strong diplomatic support. The British domestic market was now essential to Elswick but Woolwich stood in the way. Rendel challenged the government to convert Woolwich into a private company in a free market where, he predicted, it would last only a few years:

> On every ground Government orders for guns should now be thrown open. The Elswick firm has plainly out-stripped Woolwich and should have its reward. They have produced an entirely new gun of double the power of the old one, which old gun Woolwich now is forced to abandon. Suffer Woolwich to deprive Elswick of its fair reward of both credit and profit, and you inflict one more heavy blow on the private enterprise and productive power of the country. Give Elswick its due, and you check the Woolwich monopoly, encourage private enterprise generally, and give England once more the chance of recovering a fair share of the gun trade of the world, and so extending its manufacturing renown and resources.[61]

It was a blistering attack on the technological incompetence of Woolwich and the self-interested bureaucracy of the War Office, but

Elswick was coming under competitive pressure itself in the growing international markets. Krupp had benefited from the success of his field artillery in the Franco-Prussian War and was now challenging Elswick's overseas domination of the market for naval guns. At the same time, the costs of modern armaments were rising beyond the financial strength of many foreigners, a problem exacerbated by the spreading depression. Elswick had to expand to be able to manufacture the new armaments, but to survive it needed access to the British market for guns. The Royal Gun Factory stood in the way and the Woolwich steel mill became the symbol of Elswick's struggle to be once again the arsenal of Britain.[62]

The struggle began in 1884. Ground had been appropriated and cleared at Woolwich for the installation of Siemens furnaces, and Rendel wrote to Henry Brand, the Surveyor-General of Ordnance, to enlist him as an ally. The War Office strategy was to build the factory 'by stealth or surprise' before the opposition could organise itself. Rendel's great fear was that, as with guns, Woolwich would sit in judgment on the quality of its own steel and be free to reject unfairly Elswick steel.[63] A week later, Rendel sent Brand a long diatribe against Woolwich and the threatened destruction of the private trade which the proposed steel plant represented. Moreover, he questioned the ability of Woolwich to construct and operate the sophisticated technology of a Siemens furnace. A steel plant was an enormous engineering and managerial undertaking even for an experienced firm like Elswick. When they began building their steel mill in 1882, Elswick lured Sir William Siemens away from his own business with a seat on the board of directors and £30 000 in shares. Furthermore, they hired Henry Dyer from Whitworth to manage Elswick's steel works at £1500 per year plus commission. He became a director in 1886. But the War Office, said Rendel, foolishly believed that they could set up a steel works 'with nothing more than an Artillery Officer to guide them'.[64]

Brand agreed with Rendel and was working to have the private trade in Sheffield expand their steel-making capacities.[65] Those producers wanted a government guarantee of orders before they would expand, but Brand was optimistic that such an arrangement was possible.[66] An agreement was concluded and on 2 December 1884 Northbrooke, the First Lord of the Admiralty, announced that Woolwich would not build its own steel plant but buy from the private trade.[67]

But the War Office was not so easily defeated. Despite promises of two Secretaries of State for War that Woolwich would not expand its steel-making capacity beyond 6-ton ingots (used for 6-inch guns), by 1886 the Royal Gun Factory had developed the capacity to produce 15-ton ingots. The new Surveyor-General of Ordnance, William Woodall, was a prisoner of his advisors and the War Office was able to hide the expansion from parliamentary inspection. Dyer asked Rendel to take up the matter once again.[68]

Rendel wrote to Woodall who promised to obtain the 'precise conditions' previously laid down by Hartington, the Secretary of State for War, to Brand regarding the supply of steel ingots from Woolwich and the trade, and to ascertain whether or not there had been a 'breach of faith'.[69] Rendel gathered evidence and arguments with which to confront his opponents and a few days later Dyer was hopeful that 'all this stirring up' by Rendel would 'effectively stop our friends at Woolwich'.[70] Even Lord Armstrong, who disliked open confrontations with the War Office, agreed that aggressive tactics were required in the interests of Elswick.[71]

In March 1887, Dyer reported that 'the Parliamentary authorities' were 'just beginning to realize that the permanent and military authorities at the War Office have combined with Woolwich to humbug them'. The game was simple: in their Estimates, the War Office would include money to repair their 6-ton steel furnace; Parliament would pass it and Woolwich would 'repair' their 6-ton furnace into a 10-ton furnace. Later, the 10-ton furnace would be 'repaired' into a 15-ton. No one outside of the War Office would know about it until it was too late.[72] The minute from Hartington explaining government policy was conveniently lost.[73]

The office of Surveyor-General of Ordnance was a political appointment created by Cardwell in 1870 as part of his retrenchment reforms, but its control over the manufacturing departments at Woolwich was more theoretical than real. Brand and Woodall were Liberal MPs with limited knowledge of the manufacture of guns. Brand was a politician, not a weapons expert.[74] Although Woodall was a businessman with an interest in technical education, his political energies were taken up by the women's suffrage movement and after 1884 he led the parliamentary faction which supported women's suffrage.[75] It was relatively easy, then, for the Woolwich military to control their political superiors. As Dyer said, like all honest men, Woodall's weakness was his credulity: he 'naturally accepts the statements of his advisors without too close a criticism'.[76] Woolwich's claim that they could not get enough steel from the private trade was disputed by Henry Dyer after he visited all the steel plants in England. It was, he said, simply a handy argument to justify their expansion and monopoly.[77]

Woolwich used the quality control and inspection process to reject Elswick steel as defective. The procedures of the tests were confidential and Woolwich steel passed far more often than steel from Elswick or other private manufacturers. The company protested about this system, which gave the superintendent of the Royal Gun Factory 'sole authority over his own steel [,but also] uncontrolled power over the steel of all other gun manufacturers' with sufficient discretion to pass Woolwich steel ingots and reject those from Elswick whatever the test results.[78] Elswick guns were rejected with a similar sleight-of-hand. Old Armstrong guns were tested against new Woolwich guns and the results used to justify claims about the superiority of Royal Gun Factory products. Elswick was familiar with this tactic as it was a favourite technique used by Krupp.[79] However, it was apparent that the Ordnance Committee needed Armstrong's knowledge of guns and, when it sought his advice on gun construction, he demanded reforms in the system of secret and arbitrary tests.[80] Rendel and the Sheffield steel producers continued their campaign and succeeded in halting further expansion of government furnaces.[81]

In 1887, the first of a number of parliamentary inquiries began on the question of public versus private production of armaments. From the Committee on the Private Armaments Manufacturer (Morley Committee) of 1887 to the Government Factory and Workshop Committee in 1907 (Murray Committee) the trend is clear: private armaments companies (guns and ships) received a greater share of the increasing production of new naval weapons for the British Admiralty. By the 1880s, the technological gap between Elswick and Woolwich had grown so great that the Secretary of State for War, Lord Hartington, asked Armstrong and Noble to join the Ordnance Committee as associate members and attend the meetings which dealt with the construction of guns. Thus, 23 years after he had resigned from Woolwich, Armstrong once again held an official position inside the War Office.[82]

The Royal Ordnance Factory struggled to survive. Administrative reorganisations were carried out, but funds for capital investment and research and development remained inadequate. The Director-General of Ordnance, Sir William Anderson, following the recommendations of the Morley Committee, attempted to centralise and amalgamate the five manufacturing departments and streamline production, but the effort failed.[83] An attempt to create a strong design department withered in the 1890s owing to the resistance of the departmental bureaucracies which refused to relinquish any independence. Elswick, with its superior organisation, modern physical plant and high expenditures on research and development, increased its dominance over the manufacture of ordnance. Government factories degenerated rapidly in the first decade of the twentieth century. In a time of continual technological development, Woolwich managers saved their old machinery in case of emergency. In 1914, the royal arsenals had become outdated in all areas.[84]

IV

Two new powerful state players entered the debate over the merits of public or private armaments production in the 1880s. As the Great Depression persisted, the Treasury showed an interest in the matter and in 1886 agreed that the Admiralty ought to have the right to purchase guns directly from private manufacturers whenever the Admiralty were satisfied that any particular work could be done more cheaply there. Thus the Treasury reversed the old War Office argument that Woolwich acted as a check on prices charged by the trade. The Treasury hoped 'that orders would never be given by the Admiralty for the simple reason of maintaining Woolwich' and encouraged the policy of using outside firms.[85] At the same time, the Admiralty was diverging farther from its traditional policy of allowing other countries to develop new weapons and relying on British productive capacity to out-produce any competitor when war became imminent. According to this view, technological innovation in peacetime wasted money on soon-to-be obsolete weapons and stimulated naval arms races with other great powers which Britain, with the largest navy, would find the most expensive and from which she had the least to gain since she already enjoyed naval supremacy. It was for the latter reason that the Admiralty had resisted the introduction of steam power early in the nineteenth century on these grounds and it would be employed again in 1906 by those against the introduction of the *Dreadnought*.[86]

Other naval traditions also gave way in the late nineteenth century. Before 1880, many sailors still held traditional skills in high esteem and gunnery practice remained 'distinctly uninteresting' compared to the tasks of an upper yardman.[87] Before 1880, officers were reluctant to take courses in using the new naval weapons; few gunnery skills were necessary with old muzzle-loaders and hence gunnery lacked the status to attract officers.[88] After 1880, however, this was no longer the case, as guns became ever more sophisticated to fire and gunners required intensive training.

There were several organisational and historical differences between the navy and the army. The Admiralty had fewer factional divisions as its lines of authority were sharper than those in the army. An 1869 order-in-council declared the First Lord, the man who spoke for the navy in Cabinet and Parliament, alone responsible to Parliament; the other members of the Board of the Admiralty, including its chairman, the First Sea Lord, became his advisors whose only recourse against decisions with which they disagreed was to resign (the threat of which was used with effect from time to time)[89] or to manipulate the press and create a public opinion sufficient to change the minds of the First Lord and the

Cabinet (a tactic used with great effect in 1884–5). The political position of the navy vis-à-vis the army improved after 1880 for several reasons. Imperialist adventures proved more difficult, costly and controversial than in earlier times and the new high-technology 'middle class' navy was more inviting than the aristocratic army encumbered by its constitutional link with the monarchy. Naval officers were better trained and more open to technological innovations. Finally, in the 1880s, new naval officers and some politicians restored the navy's traditional reputation as the only line of home defence and protector of the Empire.[90]

The relationship between Armstrong and the British Admiralty was far more congenial than with the War Office. In 1871, the Admiralty realised that a general review of warship design and strategy had been made necessary by technological developments and a special committee on designs was created to make recommendations. George Rendel of Elswick, the architect of the *Staunch* gunboat, was a member.[91] This report reveals the continuous effort within the Admiralty to maintain a logical relationship between existing naval strategies and the constantly evolving technologies. Science was throwing up new and improved weapons; the art of the strategist was to bring into harmony the conflicting elements which made up a ship of war. The great question of the day was how engine power and sail, offensive guns and defensive armour could be united in one ship. Warship design had become the art and science of making choices:

> A perfect ship of war [the committee reported] is a desideratum which has never yet been attained, and is now further than ever removed from our reach. Any near approach to perfection in one direction inevitably brings with it disadvantages in another.[92]

By 1870, it was impossible to blend the most powerful guns with the heaviest armour without radical innovation in ship design. Most importantly, one had to choose between the great weight of offensive guns and defensive armour. The old broadside guns were designed for sailing ships; the new, heavier guns had to be on deck to maintain stability. This in turn required sails and rigging to be replaced by steam engines. And so on: one technology could only be fully exploited by sacrificing all the others. Armstrong, invited by the Admiralty to submit his views on the subject of guns and armour, argued the radical case for guns. Guns, he said, would always overtake armour and thus the latter should be wholly abandoned in favour of the former, as he had done in the *Staunch*.[93]

This was too radical for the Admiralty, but the significance of the *Staunch* could not be denied and the committee accepted the need to have a variety of warships designed for specific tasks:

> As a powerful armament, thick armour, speed and light draught cannot be combined in one ship, although all are needed for the defence of the country, there is no alternative but to give the preponderance to each in its turn amongst different classes of ships which shall mutually supplement one another.[94]

Accordingly, the Committee on Designs recommended that the *Staunch* become the nucleus of British coastal defence.[95] The first-class ironclad battleship fleets entered a transition period whereby traditional technologies (sails, broadside guns) mingled with the new (steam power, turret guns) for long-distance colonial duties. Armoured cruisers combined steam and sail and carried a mixture of heavy and light guns and some armour. The committee also concluded that 'occasions may arise in which a rapid and well-sustained fire may be of more importance than penetrative power'.[96] The Admiralty did not manufacture its guns and was not averse to developing ties with Armstrong, and when Elswick developed the quick-firing gun the Admiralty demanded the right to buy from Elswick.[97]

The technological gap between Elswick and the naval dockyards increased over the years. By the 1880s, a warship had become a vast and expensive engineering project which demanded sophisticated manufacturing organization, highly skilled engineers and knowledgable and experienced management. For example, a Committee investigating the relationship between private contractors and the royal dockyards was told by the Controller of the Navy that the dockyard engineers were not able to write down the complete specifications for a ship so that a contractor could build it. When asked why this was not possible, he explained that it would fill volumes and one 'must trust to their general knowledge of shipbuilding'. But the specifications which could not be written down were gun fittings, because guns, gun mountings and magazines were evolving even while the ship was under construction. Government naval architects could not keep up with ordnance innovations.[98]

The dockyards did not have any information about the costs of work done by the private manufacturers. Estimates for designs sent out to private firms like Elswick were 'guided in a great measure by the prices that have been offered by contractors in former cases'.[99] After a major inquiry in 1887, the Admiralty reluctantly admitted that

> the professional advisors of the Admiralty, whose duties confine them so much to office work, have probably a more limited opportunity of watching the progress of engineering science than the advisers of private shipbuilding companies.[100]

It was the committee's opinion that the Admiralty did not have 'a thoroughly practical engineering department with business capacity'.[101]

The commissioners emphasised the particular problem of engine design, but it was also generally true. The Admiralty, like the War Office, had neither the technical nor the managerial talent that Elswick enjoyed.

Elswick was also far ahead of its private competition. Other ship builders could produce hulls, but only Elswick had experience in building complete warships. In 1882, Armstrong had created an armaments-manufacturing plant of unprecedented capacity and efficiency. Elswick refined its own steel in its new, modern furnaces, built its own ships in its own shipyards and installed its own ordnance. It had become a complete arsenal, a vertically integrated armaments business able to build a warship which was ready for battle when slid into the River Tyne.[102]

This reality was reflected in the contracts given to other private firms: clauses stipulated that the turrets, guns, carriages 'and all the machinery for loading and working the guns will be supplied, under another contract, by Messrs. Sir W.G. Armstrong and Co.'.[103] Sir Edward Reed claimed that the Admiralty had lost all control over the design of its large naval guns to Armstrong.[104]

Elswick held a virtual monopoly in the production of gun carriages and the hydraulic mechanisms necessary to load and manoeuvre the giant guns. This too was suddenly discovered by the government during its investigations into the contract system. The Earl of Ravensworth, chairman of the 1884 Committee on Contracts was shocked and dismayed to learn the degree to which the British Navy relied upon Sir William Armstrong for its armaments, and its virtually total dependence upon Elswick for gun mountings. He considered it 'a serious question that the whole of our naval power in this respect should be placed under the control of one firm', especially since that firm also sold to foreign governments.[105]

In 1885, the Ravensworth Committee strongly recommended that private ship builders construct more of the navy's new ships and the royal dockyards confine themselves to repair chores and 'experimental' warships. That is, the Government would take over the less profitable aspects of the business and the most expensive costs related to research and development. New warships would be built by tender from those few private firms whose 'character and experience' were appropriate. Firms which submitted low tenders but had not the knowledge or experience to deliver on the promise would be weeded out. The committee also encouraged the navy's attempt to create its own staff of armaments designers to challenge the Elswick monopoly.[106] This effort failed: three years later, the Forwood Committee was told that Elswick was still the only source of the hydraulic machinery necessary for the heavy naval guns. All British warships required £60000 to £70000 of

Elswick's hydraulic equipment, an extremely profitable situation for the company.[107]

Rear-Admiral Thomas Brandreth rightly pointed out to the Ravensworth Committee that it was 'a very different thing building a passenger ship from building a man-of-war with the armament changing every year'.[108] Warships generated far larger profits than commercial ship building and therefore received the greatest amount of the available managerial, technological and financial resources in private firms. In 1890, a first-class merchant ship cost £28 per ton of hull and a first-class passenger ship £32 per ton, whereas a warship hull without armour cost from £60 to £65 per ton. Armour increased the cost by another £10 per ton. A battleship hull cost £80 per ton, a torpedo boat £280 per ton. The scale for engines was similar: £60 per ton for a first-class passenger ship; £85–110 for cruisers; £105 for battleships and £373 for torpedo boats.[109]

Elswick's innovation in gun technology remained strong throughout the prewar era. Woolwich finally obtained a centralised research department in 1907, but it was too little and too late to catch up to Elswick. There was no governmental design department until after the First World War,[110] and in 1910 the Admiralty relied more than ever on Elswick for technological progress.[111] Elswick warships and ordnance were considered superior to those of Krupp also[112] and Armstrong's monopoly on the production of gun-mountings remained throughout the 1890s. Armstrong faced technological competition from Whitworth on only three items in 1894 but even this threat evaporated when he bought the Whitworth firm in 1896.[113]

The growing importance of Elswick to the British navy is reflected in the Naval Defence Act of 1889. The international context of this major naval building programme is discussed in the next chapter; here the importance of Elswick technology to any British naval programme needs to be emphasised. The total spent under the Naval Defence Act was £21 500 000 for 70 ships, of which 32 were built by contract for £10 million, 38 at the government dockyards and royal factories for £11 500 000, divided into £8 650 000 for hulls and engines and £2 850 000 for armaments.[114] However, this did not mean that Woolwich received £2 850 000 in armaments orders. That amount included mountings and those came from Elswick. Similarly, some of the steel for the Woolwich guns was ordered from Armstrong and other steel companies. The productive process which lay behind British warships had become extremely complex and Elswick was in the forefront of the industry. Warships could not be built at the moment of crisis; a continuing programme was required which could incorporate the constantly developing weapons technology. The new naval armaments required an expanded volume of output to contain their spiralling costs.

Armstrong's virtual monopoly of gun-mounting technology is clearly shown by the distribution of orders in the Naval Defence Act. Elswick produced all of the gun-mountings for the first-class battleships and cruisers and the vast majority of those for lesser ships. Armstrong's orders amounted to £426 279; the Ordnance Factory and Hotchkiss received £20 341 in orders for mountings.[115] Guns for the new ships were ordered from six firms, including the Royal Ordnance Factory. Elswick received 39 per cent of the total orders (£184 971); Whitworth, still a strong competitor, received 36 per cent of the total. The Hotchkiss Company received orders for quick-firing guns which accounted for 11 per cent of the total. Vickers and Maxim received only about 3 per cent each. Overall, Elswick obtained 95 per cent of the orders for gun mountings and 39 per cent of the gun orders. Although the company was awarded only 5 per cent of the total hull production (£268 600), Armstrong was the only one of the ship builders to participate in the construction of the hulls on all four battleships built by contract. Two of the 17 cruisers were built at the Elswick shipyard.

In the 1890s, armour once again threatened to win the advantage over guns and its production expanded rapidly. Armstrong moved into that area of the armaments business when he purchased Whitworth's Openshaw plant.[116] However, the company's reputation was built on its guns and gun-mountings and the Admiralty's demand for these weapons increased dramatically after 1884. Between 1886 and 1892, the number of heavy guns in the Royal Navy increased from 499 to 1868; quick-firing guns from 33 to 1715; torpedoes from 820 to 2874.[117]

Expanded production of naval armaments in turn required a foreign policy to justify it to the British public. In the 1880s, this was not too difficult to find: threats from France and Russia were always available. It has been argued that a main reason for the Naval Defence Act of 1889 was that the 1888 naval manoeuvres indicated to the Admiralty that their fleet could not blockade the French if the latter was allied with Russia. The conclusions drawn from the manoeuvres were important factors in British policy, but the Naval Defence Act was not a decision wholly made in 1888. Hamilton, the First Lord of the Admiralty, stated, when he introduced the Bill in March 1889, that it had been the result of three years of 'investigation and work'.[118] The various committees on the army and navy estimates in 1887 and 1888 had made it clear that only by expanding the amount of work in the private sector could any naval rearmament policy be carried out. The new weapons demanded long-term planning: a battleship took three to four years to build and hence the naval defence programme could not be planned around Parliament's fiscal year. Goschen, the Chancellor of the Exchequer, stated, 'we wish to insure a regular naval programme for years to come';

that 'no change of plan on the part of the Admiralty or on the part of a succeeding Government or Parliament should cause the programme to be stopped or delayed'.[119] The four- to five-year programme could not be stopped, delayed or altered by any future (Liberal) government without an Act of Parliament. The ships were to be built as quickly as possible and with no design changes allowed during construction. Financing for the entire scheme was included in the Act and hence, once passed, was beyond parliamentary control. 'In this matter,' said Salisbury, 'we have most undoubtedly made a new departure'.[120]

Randolph Churchill in the Commons and Granville in the Lords were outraged by the scheme to place the programme beyond the authority of any future government. Granville lamented for the old policy of Peel and Disraeli which had seen diplomacy as a cheaper method than an arms race for confronting international relations.[121] The Radicals saw the Naval Defence Act as a plot to involve Britain in European power politics and said it would ignite an arms race among the great powers.[122]

Lord Charles Beresford, leading the opposition from the radical right, considered the £21 500 000 as less than half what the navy required and disagreed with Churchill's argument that the strength of the fleet should be determined by policy. Beresford had no patience for diplomacy which, he said, changed quicker than the fleet. His 'policy' was simple – if neither cheap nor practicable – 'that if we go to war with any country we should be in a position to sink every vessel of the enemy'.[123]

The Naval Defence Act is a pivot in the history of Elswick and of the British armaments industry in general. It opened an era during which national power and status were given their highest expression in the mighty ships and guns which poured out of the dockyards and factories of the world's armament companies. The technology of war continued to develop and become more destructive, as it had done since the first little Armstrong breech-loader was demonstrated in 1855 and since the *Warrior* was launched in 1860, but after 1889 a new and tighter relationship developed between those who made the tools of war and those who purchased and used them.

In the 1880s, the younger generation of British sailors, led by Admiral John Fisher, recognised the fact that naval power depended upon technological factors and that private companies like Elswick were the source of the most important innovations. As this group of officers rose to positions of greater authority, the relationship between the Admiralty and private industry matured and expanded. The traditional Admiralty policy of discouraging technological innovation gave way to an active cultivation of new weapons technologies. Admiralty men drew up plans for new armaments which they considered strategically or tactically important and discussed with the armament companies who built them the technological alternatives and the size of orders required to finance production. Designs became more sophisticated, planning time lengthened (while production time shortened) and a continuous programme of research, development and production gathered momentum.

V

A new era in naval engineering and sea power began in 1880 with two major technological developments: cheap, high-quality steel and steel alloys which gave steel whatever specific qualities were desired (tungsten and nickel for hardness, manganese for tenacity and ductility, chromium to resist rust, and so on); and, secondly, slow-burning gunpowder, which imparted a constant pressure on the shell as it travelled up the gun barrel, leaving with a greatly increased initial velocity. Nickel was a particularly important alloy for armour.[124] It is difficult to underestimate the importance of the Naval Defence Act in engineering and international history. As the Crimean War had attracted William Armstrong and many other engineers to military problems, so too the Naval Defence Act reinvigorated naval engineering and put the navy back at the centre of domestic and international affairs. Engineers were encouraged to develop more powerful and efficient marine engines, guns that could fire shells through any armour, and armour that could stop any shell. The combinations and permutations which naval engineers faced offered plenty of room for debate about what constituted the 'best' warship. The old contest between defensive armour or offensive firepower and speed became much more complicated and enmeshed in foreign policy and diplomacy. It may well be that the Naval Defence Act saved the British steel industry from German and American competition, and, given the primary role of Elswick in that revival, one might add that Elswick may have done much more for British industry in general than the anti-entrepreneurial view recognises.

Until then Armstrong, Whitworth and Krupp had been able to develop new guns and warships from their own resources. This had been a risky business since there was great resistance to them from state arsenals and old military–naval leaders who failed to grasp fully that the technological changes taking place around them were profound, that the technological innovations of private manufacturers had tactical and strategic implications and that industrial capacity had become a fundamental consideration of strategic considerations. After 1880, armaments research and production costs became extremely high. The world markets were too small to support the costs of the new industrial technologies developed by Armstrong, Krupp and the other large armaments manufacturers. The great powers had to become buyers if naval technology was to continue to develop. Furthermore, there could be no sudden withdrawal of the state from an agreement, as had happened to Armstrong in 1862. Nor could the War Office or its Board of Ordnance be allowed to continue to impede the development and deployment of innovations by private men and companies as they had done since the Crimean War.

With men like Sir John Fisher in the ascendancy at the Admiralty and the intensification of international rivalries after 1880, there was little chance of such a débâcle. The Admiralty had always been open to developing a close working relationship with private industry The construction of the *Warrior* had been a cooperative venture in which the Admiralty provided the specifications of the ship for the private builders who designed and built it. This process was revived after 1880 and became the basis of the British naval–industrial complex.[125] The system was not without conflict between technological innovations demanded by the Admiralty to satisfy the tactical and fighting requirements they deemed necessary[126] and the need for a minimum volume of orders to make the production of those innovations profitable to the private manufacturers; yet it was not so costly that government support was lost. As we shall see, this financial difficulty was partly alleviated by foreign sales of Elswick and the other private armaments manufacturers, to which the government lent its diplomatic support after 1885. The close cooperation between the Admiralty and the private manufacturers enabled naval armaments to surge forward with a wide range of technological innovations between 1884 and 1914.[127]

Between 1880 and 1914, governments redefined their relationship with private industrialists in order to maintain the facilities necessary to conduct a major war. Government arsenals and shipyards were incorporated into a larger armaments industry led by private companies whose well-being acquired a national importance. The armaments companies were the source of technological innovation and had the industrial base upon which British and European governments relied to supply the means to carry out foreign policy.[128] But on the other hand, the trend was also towards naval technology exerting an increasing influence upon foreign policy. Thus the technological rationalisation of the British armaments industry into a naval–industrial complex of which Elswick was a major part was accompanied by a close relationship between the managers of the company and the political managers of the British state.

Notes

1. Henry Guerlac, 'Vauban: The Impact of Science on War', in Peter Paret (ed.), *Makers of Modern Strategy From Machiavelli to the Nuclear Age* (Princeton: Princeton University Press, 1986), p. 68; Patrice Boussel, *Leonardo Da Vinci* (Secaucus, N.J.: Chartwell Books, 1986).
2. De Lamar Jensen, *Renaissance Europe*, 2nd edn (Lexington, Mass., D.C. Heath, 1992), pp. 213–17.
3. On the other hand, Haber's contemporary and close friend Albert Einstein eschewed the state. For a fine discussion of the two men on this point, see Fritz Stern, *Einstein's German World* (Princeton: Princeton University Press, 1999).
4. Thomas P. Hughes, 'National Socialist Ideology and German Engineers', *The Holocaust: Ideology, Bureaucracy, and Genocide*, Henry Freidlander and Sybil Milton (eds) (Millwood, New York: Kraus International Publications, 1980), pp. 165–81.
5. Mahan's *The Influence of Seapower Upon History 1660–1783* enjoyed immense worldwide popularity. See Arthur J. Marder, *The Anatomy of British Sea Power: A History of British Naval Policy in the Pre-Dreadnought Era, 1880–1905* (New York: Alfred Knopf, 1940), pp. 44–61.
6. *The Investors' Review*, 9 Dec. 1905, pp. 714–15.
7. E.L. Woodward, *Great Britain and the German Navy* (London: Frank Case, 1964), pp. 494–517.
8. 'Overloaned and Overarmed', *The Economist*, 24 May 1913, pp. 1274–5.
9. William Manchester, *The Arms of Krupp* (New York: Bantam, 1970), p. 303.
10. See Manchester, *Arms of Krupp*, 308–11 and 304–5, for a similar manoeuvre by which a Krupp agent used an invasion scare to steal a Brazilian order for the superior quick-firing guns from the French firm Schneider.
11. This tactic was no doubt borrowed from the Mulliner affair in Britain in 1908.
12. J.T.W. Newbold, *The War Trust Exposed* (London: National Labour Press, 1913), George H. Perris, *The War Traders: An Exposure* (London: National Labour Press, 1913), H.R. Murray, *Krupp's and the International Armaments Ring* (London: Holden & Hardingham, 1916).
13. Sir Edward Grey, *Twenty-Five Years 1892–1916*, vol. 1 (New York: Frederick Stokes, 1925), pp. 89–90. For a discussion of Grey's comments, see Paul Kennedy, 'Arms-races and the Causes of War, 1850–1945', *Strategy and Diplomacy, 1870–1945* (Aylesbury: Fontana Paperbacks, 1984), ch. 6. Grey concedes that Britain provoked the naval arms race by launching the first Dreadnought in 1906, but he lays ultimate blame for the war itself on Germany armaments policies. On the role of armaments in creating the 'mood' for war in 1914, see James Joll, *The Origins of the First World War* (London: Longman, 1984).
14. S.W. Roskill, *Naval Policy between the Wars*, vol. 1: *The period of Anglo-American Antagonism 1919–1929* (London: Collins, 1968). To be sure, the expense of new naval programmes and the fact that navies were so insignificant in the First World War were also good reasons for entering arms limitation agreements.

15. 'Report of the Sub-Committee of the Temporary Mixed Commission', League of Nations, A. 81 (1921), p. 11, reproduced in Noel-Baker, *The Private Manufacture of Armaments*, p. 89.
16. The name of arms firms were made famous by H.C. Engelbrecht and F.C. Hanighan, *Merchants of Death: A Study of the International Armaments Industry* (London: George Rutledge, 1934). See also George Seldes, *Iron, Blood and Profits: An Exposure of the World-Wide Munitions Racket* (New York: Harper, 1934); Philip Noel-Baker, *The Private Manufacture of Armaments* (London: Victor Gollancz, 1936). Titles from the literature of the First World War era are similarly suggestive of the conspiratorial nature of the armaments industry: J.T.W. Newbold, *The War Trust Exposed*; George H. Perris, *The War Traders*; H.R. Murray, *Krupp's and the International Armaments Ring*.
17. Clive Trebilcock, 'Legends of the British Armaments Industry, 1890–1914: A Revision', *Journal of Contemporary History* 5 (1970), pp. 3–19.
18. David Stevenson, *Armaments and the Coming of War: Europe, 1904–1914* (Oxford: Oxford University Press, 1996).
19. Clive Trebilcock, 'A "special relationship" – Government, Rearmament, and the Cordite Firms', *Economic History Review* 19 (1966), pp. 364–79; 'Legends of the British Armament Industry 1890–1914: A Revision', pp. 3–19; 'The British Armaments Industry 1890-1914: False Legend or True Utility', in Geoffrey Best and Andrew Wheatcroft (eds), *War, Economy and the Military Mind* (London: Croom Helm, 1976), pp. 89–107; *The Vickers Brothers: Armaments and Enterprise, 1854–1914* (London: Europa, 1977).
20. Robin Higham, 'Government, Companies, and National Defence: British Aeronautical Experience, 1918–1945 as the Basis for a Broad Hypothesis', *Business History Review* 39 (1965), pp. 323–47.
21. James Joll, *The Origins of the First World War*, p. 126.
22. Arthur J. Marder, *The Anatomy of British Sea Power*, p. 42. Marder had published these conclusions two years earlier in 'The English Armament Industry and Navalism in the Nineties', *Pacific Historical Review* 7 (1938), pp. 252–3.
23. *Hansard*, 18 March 1914, cols 2183–4.
24. Ibid., 13 May 1901, col. 1492. See col. 1565 for Winston Churchill's quotation from his father's 1886 resignation letter, making the same point.
25. See Armstrong's letter to Lord Dufferin of 12 April 1871 in 'Report of the Committee on Designs', *Parl. Papers* XIV (1872), pp. 521–4. Although that committee did not recommend that Britain abandon armour plating altogether, as Armstrong advocated, they did recommend the *Staunch* class of gunboats as part of 'a very valuable nucleus for a satisfactory systems of Home defence' (ibid., pp. 506, 517–18). A description and illustration of a *Staunch* is given in the *Illustrated London News* upon which Stanley F. Wright, *Hart and the Chinese Customs* (Belfast: William Mullan & Son, 1950), pp. 473–6, based his thorough description (and whose illustration he reproduced), but without acknowledgement. See also John L. Rawlinson, *China's Struggle for Naval Development, 1839–1895* (Cambridge, Mass.: Harvard University Press, 1967), p. 69.
26. Rawlinson, *China's Struggle*, p. 70.
27. T&W, 31/4619. In 1896, Armstrong designers built a 3000-ton ferry, dismantled it into 7000 parts of 20 tons or less, and sent it by the

Trans-Siberia Railway to Lake Baikal in central Russia, where it was reassembled. See Peter McKenzie, *W.G. Armstrong*, p. 102.
28. Sir William Palliser to Rendel, 8 March 1868, T&W, 31/903.
29. *Hansard*, 21 March 1859, col. 482; Armstrong evidence to the Commission on *Warlike Stores*, q. 9267.
30. *Hansard*, 24 Feb. 1862, cols 681–2; 27 June 1864, cols 336–7.
31. Robert Blake, *Disraeli* (New York: St Martin's Press, 1967), pp. 485–6. Blake points out that, in 1858–9, Disraeli, Chancellor of the Exchequer, had wanted to build cheaper frigates rather than the *Warrior* (ibid., p. 394). 'It was as if he wanted to out-Gladstone Gladstone' and 'the fact that Britannica continued to rule the waves owed nothing to Benjamin Disraeli' (ibid., p. 395).
32. C.J. Bartlett, 'The Mid-Victorian Reappraisal of Naval Policy', in K. Bourne and D.C. Watt (eds), *Studies in International History* (London: Longmans, 1967), pp. 189–208.
33. General John Adye, 'Memorandum by the Director of Artillery, on the Breech-Loading System of Field Artillery, April 1870', *Departmental Papers Relating to Muzzle-Loading versus Breech-Loading Rifled Field Guns*, W.O. 33/21A.
34. Armstrong to Rendel, T&W, 31/2270 (24 Nov. 1876); 31/2275 (19 Jan. 1877); 31/2276 (Feb. 1877).
35. See O.F.G. Hogg, *The Royal Arsenal*, vol. 2 (London: Oxford University Press, 1963), pp. 749–50, and the *World Who's Who in Science* (Chicago: Marquis Who's Who, 1968).
36. Even the great Alfred Nobel was forced to divulge details of his new compound to the War Office Committee on Smokeless Powders in order to gain admittance to the British market, and soon learned that the War Office had stolen his invention and hid behind the legal immunity of the Crown. Abel was chairman of that Committee. See W.J. Reader, *Imperial Chemical Industries: A History* (London: Oxford University Press, 1970), vol. I, pp. 140–42.
37. W.O. 32/7141.
38. For a later example, see Michael Sanderson, 'The Professor as Industrial Consultant: Oliver Arnold and the British Steel Industry, 1900–14', *Economic History Review* 31 (1978), pp. 585–600. Arnold's relationship was with Vickers.
39. Peter McKenzie, *W.G. Armstrong* (Newcastle upon Tyne: Longhirst Press, 1983), pp. 84–8. See also Sir John Adye, *Recollections of a Military Life* (1895), pp. 287–8.
40. Royal Commission on *Warlike Stores*, qs 8169–72.
41. Armstrong, 'Report on Wrought Iron Guns', W.O. 33/11. Krupp persisted with steel even after he discovered that the Bessemer steel, the rights for which he paid handsomely, was contaminated by the phosphoric ore found in the Rhur and Lorraine (to which he had access after 1871).
42. Jeanne McHugh, *Alexander Holley and the Makers of Steel* (Baltimore: Johns Hopkins University Press, 1980), pp. 108–9.
43. David Landes, *The Unbound Prometheus: Technological Change and Industrial Development in Western Europe from 1750 to the Present* (Cambridge: Cambridge University Press, 1969), p. 251.
44. J.C. Carr and W. Taplin, *History of British Steel Industry* (Cambridge, Mass.: Harvard University Press, 1962), p. 33.

45. Alan Birch, *The Economic History of the British Iron and Steel Industry 1784–1879* (London: Frank Cass, 1967), p. 376. For the growing interest of Britain's chief naval constructors in steel (in response to the French) during the 1860s and 1870s, and Bessemer's defensive retort, see Carr and Taplin, *History of British Steel*, pp. 111–12; 155–6; Duncan Burn, *The Economic History of Steel Making 1867-1939* (Cambridge: University of Cambridge Press, 1961), pp. 52–3.
46. Carr and Taplin, *History of British Steel*, p. 134.
47. Sir W.G. Armstrong, Mitchell & Col., *Prospectus* (17 November 1882), T&W, 31/5283.
48. Not including £60 000 freehold land set aside for future extensions.
49. Armstrong, Mitchell & Co., *Prospectus*.
50. David Dougan, *The Great Gun-Maker: The Story of Lord Armstrong* (Newcastle: Graham, 1971), p. 138.
51. Noble to Rendel, 25 November 1883, T&W, 31/3103.
52. Dougan, *Great Gun-Maker*, pp. 143–4.
53. See Theodore Ropp, *The Development of a Modern Navy: French Naval Policy 1871–1904* (Annopolis, Maryland: Naval Institute Press, 1987), *passim*.
54. William H. McNeill, *Pursuit of Power: Technology, Armed Force, and Society Since A.D. 1000* (Chigago: University of Chicago Press, 1982), p. 279. On the French torpedo-boat strategy, see Brian Bond, *War and Society in Europe, 1870–1970* (Leicester: Fontana, 1984), p. 53.
55. *Hansard*, 23 March 1863, col. 1765; 15 March 1866, col. 285; 24 May 1866, col. 1198.
56. *Memorandum of Agreement*, T&W, 31/868.
57. *Hansard*, 13 Aug. 1867, cols 1498–1509; 6 July 1868, cols 761–81. Between 1867 and 1878, Elswick managed to obtain only £40 840 in conversion orders. See Armstrong's evidence to the *R.C. on Warlike Stores*, q. 9473.
58. 'Report from the Select Committee on Army and Navy Estimates', *Parl. Papers* VIII (1887), Evidence, qs 7016–83, 7116–59, 7425–29.
59. Stuart Rendel, *A Memorandum Upon Gunmaking* (London, 1881). A copy is in T&W, 31/5270.
60. Ibid., p. 4.
61. Ibid., pp. 16–17.
62. Ibid., p. 18. The pamphlet was written in spring 1881, but not published at that moment because it appeared the Woolwich monopoly was about to be relaxed. Rendel waited for 18 months while Elswick negotiated the sale of its new 6-inch gun for the Admiralty. But the War Office, through its Ordnance Committee, betrayed Armstrong, and thus Rendel published his pamphlet.
63. Rendel to Brand, 30 September 1884, T&W, 31/3112. See also Dyer to Rendel, 13 March 1884, T&W, 31/3119.
64. Rendel to Brand, 5 October 1884, T&W, 31/3115.
65. Of whom Vickers was the most important.
66. Brand to Rendel, 4 October 1884, T&W, 31/3113 and 6 October 31/3114.
67. *Hansard*, 2 Dec. 1884, cols 429–30.
68. Dyer to Rendel, 9 Nov. 1886, T&W, 31/3125.
69. Woodall to Rendel, 11 Nov. 1886, T&W, 31/3128.
70. Dyer to Rendel, 13 Nov. 1886, T&W, 31/3130.

71. Armstrong to Rendel, 11 Oct. 1884, T&W, 31/3116.
72. Dyer to Rendel, 9 March 1887, T&W, 31/2884. Dyer had known of Woolwich plans to defy their political supervisors in late 1885. See Dyer to Rendel, 20 Dec. 1885, T&W, 31/2681 and 30 June 1886, 31/3111.
73. Dyer to Rendel, 1 March 1887, T&W, 31/2873.
74. *D.N.B.* (1906).
75. *D.N.B.* (1901).
76. Dyer to Rendel, 9 Nov. 1886, T&W, 31/3125.
77. Dyer to Rendel, 30 June 1885, T&W, 31/3111 and 20 Dec. 1885, T&W, 31/2681.
78. Elswick to the Undersecretary of State for War, 12 Sept. 1883, T&W, 31/3102.
79. Noble to Major G.H. Ellis (Secretary of the Ordnance Committee), 3 Nov. 1883, T&W, 31/3348.
80. 'Recommendations of the Ordnance Committee (with Special Associated Members) as to the Construction of Ordnance', *Parl. Papers* XIV (1884–5), pp. 3–4.
81. *Hansard*, 21 Feb. 1887, cols 177–8 and 8 March 1887, cols 1583–4.
82. 'Recommendations of the Ordnance Committee (with Special Associated Members) as to the Construction of Ordnance', *Parl. Papers* XIV (1884–5). The date of Armstrong's appointment to this committee is revealed in *Hansard*, 24 Aug. 1886, cols 373–4.
83. The Gun Factory, the Carriage Factory, the Royal Laboratory (ammunition), Waltham Abbey (gunpowder) and the Enfield and Birmingham Small Arms Factories made up the government's arsenals.
84. Great Britain, Ministry of Munitions, *History of the Ministry of Munitions* (London H.M.S.O., 1922), vol. VIII, pt 1, pp. 3–13.
85. 'Record of the Second Meeting held at the War Office by the Inter-Departmental Committees', 18 October 1886, in 'Report of the Committee on Naval Estimates', Appendix XI, *Parl. Papers*, VIII (1887), p. 767.
86. James P. Baxter, *The Introduction of the Ironclad Warship* (Cambridge, Mass.: Harvard University Press, 1933), p. 116.
87. Sir Seymour Fortescue, *Looking Back* (London, 1920), pp. 25–6. As recently as 1942, old sailors argued that naval armaments manufacturers ought to consult sailors before proceeding with any new design. See Lord Chatfield, *The Navy and Defence: The Autobiography of Admiral of the Fleet Lord Chatfield* (London: Heinemann, 1942), pp. 53, 169–70.
88. 'Report of the Committee on the Advanced Class of Artillery Officers', *Parl. Papers* XIV (1872).
89. Lord George Hamilton, *Parliamentary Reminiscences and Reflections 1886–1906* (London: Murray, 1922), p. 98; C.J. Bartlett, 'The Mid-Victorian Reappraisal of Naval Policy', in K. Bourne and D.C. Watt (eds), *Studies in International History*, pp. 197–8.
90. Franklyn A. Johnson, *Defence by Committee: The British Committee of Imperial Defence 1885–1959* (London: Oxford University Press, 1960), p. 15.
91. 'Report of the Committee Appointed by the Lords of the Admiralty to Examine the Designs Upon Which Ships of War Have Recently Been Constructed', *Parl. Papers* XIV (1872).
92. Ibid., p. 508.

93. W.G. Armstrong to Lord Dufferin, 12 April 1871, ibid., pp. 521–6. Sir William had just acquired the rights to make and sell naval Gatling guns and he therefore took this opportunity to recommend those.
94. 'Report, Committee on Designs', *Parl. Papers* XIV (1872), p. 517.
95. Ibid., pp. 517–18.
96. Ibid., p. 519.
97. *Hansard*, 7 Sept. 1886, col. 1515.
98. 'Report of the Committee Appointed to Inquire into the Conditions Under Which Contracts are Invited for the Building or Repairing of Ships' (Ravensworth Committee), *Parl. Papers* XIV (1884–5), Evidence, qs 116–25, 803–6.
99. Ibid., q. 126.
100. 'Report of the Committee Appointed by the Lords Commissioners of the Admiralty to Inquire into the System of Purchase and Contract in the Navy' (Forwood Committee), *Parl. Papers* XVI (1887), p. 539.
101. Ibid.
102. A detailed description of the Elswick Works is in the special edition of the *Pall Mall Gazette 'Extra' No. 15* (1884). See also 'Plan of the Elswick Works', Armstrong Archives, T&W (no number).
103. Ravensworth Committee, Appendix 13.
104. Ibid., qs 1844, 1890.
105. Ibid., q. 2148.
106. Ibid., pp. 130–31.
107. Forwood Committee, qs 3518–26.
108. Ravensworth Committee, q. 122.
109. Nathaniel Barnaby, 'Shipbuilding', *Encyclopaedia Britannica*, 9th edn (1890), p. 821.
110. Hogg, *Royal Arsenal*, vol. 2, pp. 1116–17.
111. Falkner to Rendel, 19 Aug. 1910, T&W, 31/7604.
112. Demetrius C. Boulger, *The Life of Sir Halliday Macartney* (London: J. Lane, 1908), pp. 445–6.
113. Admiral Fisher to Lord Spencer, 22 October 1894, Arthur Marder (ed.), *Fear God and Dread Nought: The Correspondence of Admiral of the Fleet Lord Fisher of Kilverstone* (London: Jonathan Cape, 1952), vol. 1, p. 123.
114. 'Navy Shipbuilding Programme, 1889–90', *Hansard*, 3rd series, 333 (1889), Appendix.
115. These and the following figures are calculated from 'Naval Defence Act, 1889: Contracts Entered into by the Admiralty by Virtue of the Act', *Parl. Papers* XLVI (1890), pp. 642–3; LI (1892), p. 573.
116. *Hansard*, 5 July 1901, col. 1022.
117. Oscar Parkes, *British Battleships, "Warrior" 1860 to "Vanguard" 1950: A History of Design, Construction and Armaments* (London: Seeley Society, 1966), p. 352.
118. *Hansard*, 7 March 1889, cols 1173–4.
119. Ibid., cols. 1195–6.
120. Ibid., 27 May 1889, col. 1061.
121. *Hansard*, 7 March 1889, col. 1194; 27 May col. 1069.
122. See speeches by Labouchere and Fowler, *Hansard*, 6 May 1889, cols 1313–14, 1329, and Campbell-Bannerman, 1 April, cols 1274–9.
123. *Hansard*, 7 March 1889.

124. J. Barba, 'Recent Improvements in Armour Plate for Ships', *Transactions of the Institute of Naval Architects* 32 (1891), pp. 140, 152–3; William Beardmore, 'Nickel Steel as an Improved Material for Boiler Shell Plates, Forging, and Other Purposes', *Transactions of the Institute of Naval Architects* 38 (1897), p. 276; H. Garbett, *Naval Gunnery: A Description and History of the Fighting Equipment of a Man-of-War* (London: George Bell & Sons, 1897); H.W. Wilson, *Ironclads in Action* (London, 1896); William Manchester, *The Arms of Krupp 1587–1968* (New York: Bantum, 1970), p. 242; John Perkins, 'Coins for Conflict: Nickel and the Axis, 1933–1945', *The Historian* 55 (1992), pp. 86–9.
125. This process is called 'command technology' by William H. McNeill. See *The Pursuit of Power*, pp. 278–80; Clive Trebilcock writes about 'A "Special Relationship" – Government and the Cordite Firms', *Economic History Review* 19 (1966), pp. 364–79.
126. McNeill's 'command technology'.
127. Hugh Lyon, 'The Relations Between the Admiralty and Private Industry in the Development of Warships', Brian Ranft (ed.), *Technical Change and British Naval Policy, 1860–1939* (London: Hodder and Stoughton, 1977), pp. 37–64. This article shows clearly that the source of innovation remained within the private industry.
128. Maurice Pearton, *The Knowledgeable State: Diplomacy, War and Technology Since 1830* (London: Burnett Books, 1982), pts I–III.

CHAPTER EIGHT

The political–industrial complex

Introduction

Nowhere were the financial retrenchment policies of both the Gladstone and Disraeli cabinets applied more stringently than against the British navy between 1863 and 1884. They viewed the new steam-and-iron navy as an extravagant expense with no clear strategic purpose since Armstrong's monster guns in Palmerston's great coastal forts had shifted responsibility for home defence from the navy to the army. The remaking of the British navy slowed considerably as naval estimates were reduced, innovative private entrepreneurs like Armstrong and Whitworth were ignored and the managers of the government arsenals and dockyards were denied the resources to pursue new technologies.

During the 1880s, however, home defence was linked to imperial defence, an idea which brought the navy back to the centre of British political life. The return of naval strategy changed the relationship between the British government and Elswick in two areas. First, as we have seen, a rational division of labour was developed between the private and public arsenals and dockyards, along with greater collaboration on technical questions. The second area of cooperation between the government and Elswick, the subject of this chapter, was between the managers of the company and the managers of the British state at the highest levels. If the former can be referred to as a 'naval–industrial complex', the latter might appropriately be called a 'political–industrial complex'. The merging of Elswick with the government in the 1880–1914 period was part of much larger historical developments of the time. The world changed rapidly and profoundly after 1880 and Elswick found itself at the centre of an international armaments industry of unprecedented dimensions and a key factor in the calculations of the British government. This part of the Elswick story will thus be told in the context of world events, a period during which internal technological innovations and the forces external to the company drove demand for armaments to unimagined heights. New domestic political pressures and economic realities combined with technological leaps created uncertainties, anxieties and new opportunities in the struggle for political power at home and an improved (or impoverished) position in the global balance of power.

Between 1880 and 1914, the world's great powers mobilised their social, economic, political and ideological forces for war. Admirals and generals built up their military and naval forces while politicians reconstructed their diplomatic alignments on the assumption that the practice of international relations was mostly a matter of ensuring that, when war came, their country was in a strong military and naval position. Few doubted the inevitability of war and many looked upon it as beneficial and even necessary.

I

The British navy fell out of political favour and the popular mind after the death of Palmerston in 1865. There were good reasons for complacency: the industrialisation and mechanisation of naval power, begun by the French in the 1840s as a means to overtake British naval superiority, only widened that gap with the building of the *Warrior* class in the 1860s. Indeed, it was the Royal Navy which held the advantage based, not on numbers, but on technological superiority, the very objective that French planners had pursued since Joinville wrote his pamphlet. When Napoleon III, the most ardent advocate of naval superiority through technological innovation, realised that the fewer British *Warrior* class of warships far outclassed the French ironclad *La Gloire* class in fighting power, he lost interest in continuing the naval arms race with England and turned his attention elsewhere for a foreign success.[1] The defeat of France in 1870–71, the fall of Emperor Napoleon III and the establishment of the Third Republic suggested that any French threat had been virtually eliminated for the foreseeable future. Russia had taken advantage of the Franco-Prussian War to abrogate the humiliating clauses in the 1856 Treaty of Paris neutralising the Black Sea, but still had no fleet to actually station there, an absence for which Russia suffered in its war against Turkey in 1877.[2] British defence complacency was reinforced by Armstrong monster guns housed in Palmerston's forts and carried on board British ironclads, the penetrating power of which continued to increase their advantage over armour.

Domestic issues rose up the parliamentary agenda: electoral reform, social legislation and Irish Home Rule dominated British politics from 1865 to 1884 as Britain made what Lord Derby called the 'leap in the dark' into mass democracy of universal male suffrage. From the War Office, Edward Cardwell centralised the army at the top under the control of the secretary of state by placing the commander-in-chief and the part-time militia in the counties under the direct control of the secretary of war, modernised it by abolishing the purchase of commissions, and reformed it at the bottom by abolishing flogging and reducing length of service. Even Gladstone was so impressed by the Prussian army's victories in France in 1870 that he wrote an (anonymous) article for the *Edinburgh Review* on the subject of military efficiency. On the other hand, neither Gladstone nor Disraeli concerned themselves about modernising the navy. In 1871, the army budget was increased by £3 million to a total of £16 million, but the navy's expenditures rose only by £400 000. Ships still seemed a waste of money when it came to warfare. The surprising speed with which the Prussian army overwhelmed the French in 1870, even capturing the emperor at Sedan, prompted all states to increase sharply their military spending. In Britain, Cardwell worked to make the army more efficient and more appealing to the officers and infantry soldiers. The navy was ignored. British defence strategy continued to rest on the recommendations of the 1859 Royal Commission on the Defence of the United Kingdom, which shifted the defence of the home island from the navy to Palmerston's coastal forts, or more accurately to their Armstrong monster guns (built by Woolwich Arsenal, not Elswick). The navy's visible mission was reduced to transporting the army to their colonial stations. In 1867, John Colomb, a captain in the Royal Marines, contested the coastal defence policy as inappropriate for a world trading empire, and urged that defending Britain meant protecting the sea lanes which connected Britain to her empire. Britain's true interests were imperial and only a strong navy could defend them.[3] Gladstone was a radical on armaments and empire, and even the great imperialist Disraeli saw no reason to finance the modernisation or enlargement of Britain's fleets.[4]

But the revolution in naval technology which began in the 1840s accelerated in the 1860s and 1870s, driven by the basic battle between guns and armour and, one might say, between the gun makers Sir William Armstrong and Sir Joseph Whitworth, on the one hand, and the Admiralty's chief constructors, Edward Reed and Nathaniel Barnaby, on the other. British politicians and members of the public may have ignored the navy, or felt complacent or, if radical, considered it a relic of the warlike past, but inside the Admiralty the two decades after the Crimea were very busy with attempts to resolve the technological dilemmas of battleship design.[5]

Battleship defence was reborn in Italy. As diplomatic relations with France deteriorated in the 1870s, a naval engineer with left-wing political views, Admiral Benedetto Brin, and the minister of marine from the extreme right, Simone Pacoret de Saint Bon, overcame their political differences by achieving complete agreement about the design and strategy for a new Italian navy, and together they laid the technological and political foundations for the modern Italian navy between 1872 and 1877.[6] Armstrong played a leading role in helping Brin found Italy's warship-building industry by building a gun factory at Pozzuoli, on the Bay of Naples. One can dispute whether or not this was a rational or long-term profitable investment for the company, but there is no doubt that Pozzuoli was an essential part of Italy's industrial take-off in the late nineteenth century.[7] Those who like to measure entrepreneurship should include the extent to which Armstrong contributed to the industrialisation of other countries, and ask how much that global industrialisation contributed to the economic growth of Britain. Nineteenth-century British entrepreneurship must be tested in its global context.

As with any state, Italy's strategic situation involved geography. Italy's many coastal cities were vulnerable to bombardment from battleships and to amphibious attacks. Just as France had struggled for naval advantage over England through technological innovation, Italy could not match France in numbers of ships, hence Brin's strategy of building ships that were technologically superior to the French, in this case in brute power: that is, ships made on the Armstrong principle of combining the biggest guns with the greatest speed, adjusting the heaviest armour to those two main objectives. Brin's warship was designed to carry Armstrong's monster guns which Elswick built for his two new ships, the *Duilio* and the *Dandola*.[8] These two ships were faster and carried heavier guns than any ship in the French or British navies. Armstrong was also developing an unarmoured cruiser along the same principle, the first built and sold to Chile in 1883. The *Esmeralda*, too, was far ahead of British cruisers in speed and firepower, and could, said Armstrong, take on the great lumbering battleships that England continued to build. The chief constructor of the British Admiralty, Sir Edward Reed, was an adamant proponent of heavy armour and refused to sacrifice protection to ordnance and speed, one good reason why Armstrong did very little business with the British Admiralty. The technological battle between guns and armour was won by guns, and the engineers' task was now to find the optimum combination between size of guns, thickness of armour and speed – in other words, a technological struggle between Sir William Armstrong and the British Admiralty.

The breakdown of Italo-Franco relations also brought the French navy back to life. After stagnating in the aftermath of the defeat of 1871, the French navy launched a renewal of its Mediterranean fleet in 1878. When Russia began a similar expansion of the Russian Black Sea fleet, the possibility of a Franco-Russian naval alliance in the Mediterranean prompted concern among British naval authorities about the ability of the Royal Navy to defend the entrance to the Suez Canal, Britain's trade artery to India and Asia, a main market for British exports. The bombardment of Alexandria in the occupation of Egypt in 1882 gave the Royal Navy a new respectability, and the question of naval power came forward with French and German colonial annexations in 1884–5. Imperialism and navies fed upon each other and contributed to the context for the transformation of Elswick into a giant industrial enterprise. Later, as that armaments business achieved a certain size, a third industrial factor added to the momentum towards a global naval arms trade. If imperialist conquest required naval power, imperialist rivalries amongst the great powers required more naval power and, when the great powers had completed carving up the world into imperialist spheres, naval power continued as an arbitrator of great power relations. In 1884, the balance of naval power became a political issue for the first time since 1859. It involved political agitation from both the governing Liberal Party and the Conservative Opposition.

War between Russia and England seemed imminent immediately after the end of the Russio-Turkish war of 1877–8. The warlike emotions of the British public which had been primed in 1876 by Gladstone's best-selling pamphlet against Turkey (which was of course actually directed against Disraeli), were reignited in 1878 by Disraeli's rhetoric against Russia in February 1878. The crisis lasted several months, during which 'jingoist' songs prepared the population for war. War was avoided since neither Russia nor Disraeli was prepared for one, perhaps because the objective was unclear, perhaps out of fear of repeating the mutual disaster that occurred in the Crimea in 1854. Nevertheless, Disraeli was especially drawn to the eastern Mediterranean as a vulnerable point along England's trade route to India, acquiring the Suez Canal in 1875 and Cyprus in 1878, in between appointing Queen Victoria as Empress of India in 1876 to add substance to his imperialist notions outlined in his Crystal Palace speech of 1872, and implementing a 'forward' foreign policy against Russia in the Persian Gulf region and against the Zulu and Boer states in South Africa.

All of these imperialist innovations delivered a harvest of political support for the Conservative Party, and even gave some young Liberals something to think about. All in all, the stage was set for a resurgence of British naval power. This final step in launching the age of imperialism and naval expansion required a Royal Commission to overturn the Liberal (or at least Palmerstonian) defence strategy based on landed coastal defence of the home islands, and endorse Colomb's claim that British national security included an imperial and world trade dimension. Hence the Royal Commission of 1879 was on the 'Defence of British Possessions and Commerce Abroad'. Naturally, it was chaired by Disraeli's colonial secretary, Lord Carnarvon. Once its recommendations were released, agitation could begin on behalf of a new, modern and imperial Royal Navy. Disraeli and the Conservative Party lost the 1880 election which brought back Gladstone, but the anti-imperialist

'Little Englander' could only slightly delay naval expansion. Indeed, his bombardment and occupation of Egypt in 1882 increased demands for an imperialist policy and placed the navy at the centre of the electoral struggle for power.

Inside the Liberal Party the main obstacle to naval expansion was Gladstone. International relations were changing rapidly in the early 1880s and many young Liberals had become restless with Gladstone's obsession with Ireland. Even Bismarck entered the imperialist age, proclaiming protectorates at Angra Pequena, Togoland and the Cameroons in July 1884 and annexing New Guinea. Indeed, it was to Berlin, not London, that the great powers sent their representatives in November 1884 to settle the various matters associated with the partition of Africa. The conference met until February 1885, and Germany thereupon annexed East Africa.

The Admiralty grew impatient with the technological stagnation of their fleets, especially for the new steel breech-loaded guns that Armstrong was now manufacturing and selling abroad. H.O. Arnold-Forster, political editor of *The Statist* and a Liberal candidate for Devonport, whose articles in *Nineteenth Century* on the inadequacies of the navy's ships and armament had no effect on the government, decided to appeal to a wider public through the popular press. Arnold-Forster was one of many navalists who built substantial political and newspaper careers by making the state of Britain's navy a point of political importance, insisting that British sea power was no longer sufficient to defend the homeland and its worldwide imperial interests. He was an active member of the British Navy League, founded in the wake of the spirited estimates debate of 1893–4. As the Admiralty parliamentary secretary, he returned from a visit to German dockyards in 1902 with alarming news of German naval expansion. Here, too, he implied that the Admiralty was failing in its responsibilities. He was put in charge of the War Office the next year, where he hounded Balfour for a statement defining the purpose of the army.[9] He approached the editor of *The Pall Mall Gazette*, W.T. Stead, to see if he would take up the issue of the state of Britain's navy. Stead, who had succeeded John Morley as editor in 1883, agreed, publishing one of Arnold-Forster's articles on the first page of the *Gazette*'s 14 August edition.[10] Stead was energetic, enthusiastic and the most innovative practitioner of what Matthew Arnold called the 'New Journalism' of the headline, the editorial, the personal interview and the investigating, crusading reporter writing the sensational exposé. Thus Stead leapt from uncovering the 'Truth' about the navy to uncovering a white slave trade in London, attracting more readers and increasing the political impact of the paper.[11]

There was much discontent at the Admiralty and, through Arnold-

Forster, Stead gained an interview with the First Sea Lord, Sir Astley Cooper Key, and Admiral Beachamp Seymour, both of whom had tried but failed to convince Gladstone that the navy's budget was dangerously low. But, said Key, Gladstone's mind remained focused only on Ireland and home affairs. Stead interviewed a number of Admiralty officials, admirals and commanders, who provided him with more facts about the navy, probably on the orders of Admiral Sir Geoffrey Phipps-Hornby, one of the most active proponents on behalf of a larger navy to defend colonial commerce as well as the Channel.[12] It was through this innovative investigative work that Stead was introduced to the energetic and very ambitious captain of the *Excellent*, the navy's gunnery school, Captain John Fisher. The two joined forces: Fisher provided the facts, Stead the publicity on behalf of more money for the navy. These three men of politics, the navy and the press worked together to create a public opinion which would force Gladstone to modernise the British navy with the newest technology and techniques.

With his research completed, Stead began a series of articles in the *Pall Mall Gazette* in September 1884, the first entitled 'The Truth About the Navy' by 'One Who Knows', in which he argued that, as a result of the recent scramble for colonies, the Royal Navy was no longer able to protect British interests (now defined as global) and required a massive infusion of money to bring it to its rightful state as the world's most powerful naval force. In a style and form for easy reading, Stead presented a mass of statistical information which compared the size, age, speed, armour and guns of British and foreign warships, and noted the rising naval expenditures of European states (especially France), with the drop in spending by recent British governments. The campaign had its effect as discontent spread inside and outside the navy over the belief that the lack of government funding placed England's security in jeopardy.[13]

Conservative MPs found they had an audience on the naval issue. W.H. Smith and Sir Edward Reed, the former chief constructor, contributed articles to the *Pall Mall Gazette* with war scares and predictions of disaster.[14] Lord Henry Lennox, a former secretary to the Admiralty in the Derby–Disraeli governments of 1866–8, was the Conservative's leading naval spokesman in the House of Lords, where Gladstone's First Lord of the Admiralty, Lord Northbrook, also sat, but Stead wrote that the naval question was too serious a matter to be determined by party politics.[15] Gladstone's Midlothian campaign of 1880 had demonstrated to the Conservatives the electoral power of making foreign policy a popular political issue, and the extension of the franchise in 1884 encouraged the Conservatives to engage in popular politics, and beat Gladstone at the game he invented. It was a party political struggle.

At the annual meeting of the company, Armstrong, whose family background was liberal, indeed radical, and whose armaments business was established under the first Liberal government in 1859, expressed his surprise that Britain's unprotected commerce had been regarded with so much apathy.[16]

Lord Carnarvon's aim in his 1879 Royal Commission to publicise the need to strengthen the navy, to protect overseas trade in time of war, was frustrated by the victory in the 1880 general election of Gladstone, whose government suppressed the report by expanding the promise of secrecy to a few witnesses to cover the whole report, and by printing only a dozen copies.[17] Stead's articles had generated some interest in London, but would fade away without a political campaign by the Conservatives. W.H. Smith, the First Lord of the Admiralty in the previous government, called for a commission of inquiry, but that too failed to generate any political pressure on the government, or even much interest in Parliament. Edward Hamilton, Gladstone's private secretary, wrote that the December 1884 naval debate 'was languid and the House of Commons thin, notwithstanding all the boasted interest taken in the state of the Navy and all the lashing of the country into excitement by the Pall Mall Gazette'.[18]

Conservatives persisted in their attempts to generate interest in the navy. A public meeting was held at the Cannon-street Hotel on 16 April 1885, organised by what *The Times* called

> many influential and leading men of business, without reference to party politics ... for the purpose, according to the meeting organizers, to give expression to the great anxiety that is universally felt at the present state of the Navy in its effect both upon the welfare and existence of the Empire, and its injurious and paralysing influence upon commerce.[19]

With the exception of the liberal–imperialist W.E. Forster, however, the men on the stage who spoke were predominantly influential and leading men of the Conservative political party and of the Admiralty (or both): W.H. Smith, Sir Edward Reed, Sir John Hay, Lord Eustace Cecil and a number of others. Speeches were delivered amidst cheers about Britain's dependence upon overseas sources for food and the French threat to proclaim food a non-neutral commodity during wartime, insurance rates on merchant ships and, from Sir John Hay, on the aversion of the Admiralty 'to attend to a thing which everyone clearly knows is a necessity of modern warfare – that is, speed', and on its refusal to protect their ships against torpedoes, and its 'stripping of our great ships of their armour'. He failed to mention that the way to increase speed was to reduce the amount of armour. W.E. Forster, a Liberal, was loudly cheered. He insisted that the gathering was not a party meeting, and he would not have attended if it had been called to attack the government or the Admiralty: 'They may be to blame. I do not say whether they are or not.' He spoke, as Malmesbury had in 1859, of a great war and great sacrifices which England might be called upon to make and, echoing Disraeli to loud cheers, avoided attacking the government: 'But it is just the time in which, unless we rise up as one man to displace the Government, we must have confidence in them that they will do what they can to preserve "peace with honour".' The purpose of the meeting, Forster imagined, was to strengthen, not weaken, the hands of the government.

On 28 May, the London Chamber of Commerce held a public meeting on the navy, with Phipps-Hornby delivering an address on the defence of merchant ships in wartime. This gathering was attended by military, naval and commercial leaders, including the Liberal Thomas Brassey. The naval campaign was gathering momentum, the Cabinet came under pressure and Gladstone was finally forced to relent.[20] The Admiralty was given an extra £5 525 000 to spend.[21] In June, the Conservatives came to power under Lord Salisbury, and British foreign policy moved towards the Palmerstonian emphasis on linking diplomacy more directly with armed force in the defence of British global interests and prestige. Like Gladstone, Salisbury preferred the system of international law and the Concert of Europe, but, unlike his predecessor, he was willing to spend whatever was required to modernise the British navy and ensure its superiority where it mattered, and in the 1880s that was in the Mediterranean.

Political agitation on behalf of the navy had subsided only for a short time when events in Europe generated further anxieties about the ability of the Royal Navy to protect the British position in Egypt.[22] Even the Channel seemed vulnerable to French cruisers and torpedo-boats, whose leading advocate, Admiral Aube, had entered the French Cabinet. The emergence of French revanchism and the rise of General Georges Boulanger exacerbated the fear and uncertainty in England – a *coup* was expected in France in January 1889 – as did the growing friendship between France and Russia, the third-largest naval power in Europe and a substantial naval force in Asia. Russia too had adopted cruiser warfare as a less expensive alternative to battleships. In 1888, the French mobilised their cruisers and ironclads at Toulon in order to intimidate Italy, with whom relations had been deteriorating since France had moved into Libya in 1882 and Italy had signed the Triple Alliance in 1883.

Franco-Italian tensions entered a period of crisis after the Mediterranean naval agreements of 1887 brought Italy closer to England and the two countries seemed on the verge of war. Many in England, however, thought that the French were preparing for war against them and, fearing a Franco-Russian alliance, made demands for an expanded British navy; these cries grew louder especially within the Conservative Party and among its business and press supporters. The Admiralty insisted that the navy was prepared to defend British interests, but a report of three admirals on the naval manoeuvres of 1888 concluded that the navy had lost its ability to blockade (French) cruisers effectively. These political and strategic pressures increased and produced the passage of a massive five-year building programme carried out under the Conservative government: the Naval Defence Act of 1889 which, as we saw in the previous chapter, depended to a large extent upon Elswick's technological innovations and industrial capacity.[23]

Between 1884 and 1893, public opinion was mobilised on behalf of enlarging and modernising the Royal Navy, the British part in a remarkable global phenomenon. The newspaper and political campaign of 1884–5 produced the Northbrooke programme, and that initial success spurred on the navalists. Arousing and maintaining public interest in the navy proved very successful in raising the naval estimates. In 1887, the fiftieth anniversary of Queen Victoria's coronation, the Admiralty put on a naval jubilee at Spithead which cemented the monarchy, the navy and the empire as the new sense of national identity in the public's consciousness. The event prompted much press discussion of the strengths and weaknesses of the navy, now a regular topic of public discourse. Newspapers began to record the most obscure and trivial information on naval affairs in new daily Naval Intelligence columns, along with lengthy articles by special correspondents.

In 1888–9, naval scares proved very effective in the passing of the Naval Defence Act which began the naval arms races throughout the world. In the 1890s, the navy embarked on 'hurrah trips', coastal cruises to show off its impressive ships; in 1891, a six-month naval exhibition attracted two and a half million visitors, further embedding the navy in the public mind. Between 1890 and 1895, W. Laird Clowes, the naval correspondent for *The Times*, published his seven-volume history of the navy, and Alfred Mahan of the American Navy published naval histories on the vital importance of command of the sea. Mahan's books were an overwhelming success amongst naval experts and public alike, and he was given the royal treatment on his visit to England in 1894. In 1893, the Naval Records Society was founded to publish naval biographies and histories, and in 1894 the Navy League was founded as a political lobby for a big navy by promoting fears of 'starvation', on the one hand, and Trafalgar Day on the other. In 1891, Sir Charles Dilke and Spenser Wilkinson published *Imperial Defence*, which argued Colomb's point that British imperial and national interests were one and the same, and that only sea power could protect both. All in all, the navy soared to the top of the political agenda.[24]

The 1892–3 naval scare mobilised in reaction to the Franco-Russian Alliance was much more widespread than its predecessors and represents the grand climax of the Conservative campaign, in that it brought about Gladstone's final resignation. In 1893, an even larger naval budget was passed by the Liberal government. Fears of a Franco-Russian armada in the Mediterranean set the stage for another newspaper exposé, this time a description by *The Times* of the dockyards and naval preparations at Toulon. Combined with the presence of Russian squadrons in the Mediterranean, a torrent of public debate was unleashed over the relative strengths of the British, French and Russian navies, and the demands for a vast increase in the British Navy from Liberal, Conservatives and Radicals overwhelmed the intransigent but old and ailing Gladstone. On 1 March 1894, he resigned and the Liberal imperialists took over the Cabinet; then, on 12 March, the First Lord of the Admiralty, the Earl Spenser, announced his five-year naval building programme. Expenditures for the period 1894–5 were more than £17 million, the largest amount ever spent on the British navy.[25]

Naval power came to be seen as the most important force in global affairs, the force that determined the outcome of the Sino-Japanese War in 1894–5, the Spanish–American War, the Fashoda Crisis and, most importantly, the Boer War of 1899–1903. The Anglo-Japanese Alliance of 1902 further enhanced the reputation of the navy as a vital force. The crushing naval defeat of Russia by Japan was the highlight of the Russo-Japanese War of 1904–5, not the equally impressive and strategically more important military victories of Japan along the Yalu River.

Shifting strategic naval balances continued to affect British diplomatic goals, and indeed brought her out of a long period of diplomatic independence from the other great powers of the world. British diplomacy adjusted to the emergence of two new naval powers in the 1890s, the United States and Japan. The Hay–Pauncefote Treaty of 1901 began a new era of good relations with the United States, and the Royal Navy withdrew from that part of the world. Japan developed into an Asian power by building a high seas fleet – many ships of which were made at Elswick – and the Anglo-Japanese Treaty of 1902 eliminated Japan as a Russian ally in any war between Britain and Russia. After 1904, Britain drew closer to France as the rapidly-growing German navy challenged British naval supremacy. Between 1904 and 1914, a further series of diplomatic crises and regional wars stimulated governments to increase the size, power and speed of their navies. Although the great naval building of Britain and Germany between 1906 and 1914 is the best-known naval armaments race, the phenomenon was global, with several lesser regional naval build-ups. As naval power rose, so too did Elswick, a main source of naval weapons.[26]

II

The British economic dilemma was also conducive to an expansion of the Royal Navy after 1880. Demands for an expanded navy emerged amidst a worldwide depression which left British merchant shipyards idle. The Third Reform Act of 1884 increased the representation of the large cities where industrial unemployment was concentrated, and the pressure grew to maintain profits and employment in British shipyards by increasing the demand for British warships, one way in which a Victorian government could directly affect economic development without creating dangerous ideological reverberations. Introducing his supplemental naval estimate in 1884, the First Lord of the Admiralty (Northbrook) pointed out:

> if we are to spend money on the increase of the Navy, it is desirable, in consequence of the stagnation in the great shipbuilding yards of the country, that the extra expenditure should go, not to add to our permanent staffs, but to increase the work by contract in the private yards.[27]

Northbrook's Conservative successor recalled how the Reform Act of 1884 'swamped the old niggardly and skinflint policy of the Manchester School' and that 'the mass of the recently enfranchised' were 'very proud of their Navy'.[28]

Thus the quest for political advantage, popular support, economic problems, pressure groups inside and outside the Admiralty, international affairs and the growth of foreign navies pushed British governments into a vigorous expansion of the navy between 1884 and 1914.[29] In the mid-1890s, the naval estimates, for the first time, exceeded those of the army. Because no one could argue in 1898 that the British navy was anything but supreme, the climax of Anglo-French antagonisms at Fashoda did not produce a naval scare in England. Indeed, the French worried about a preventive war by the British who were no longer concerned about a Franco-Russian naval combination in the Mediterranean. That the crisis was resolved by a French withdrawal confirmed for many in England that their absolute naval supremacy had saved them from war and reinforced the belief that it was through sea power, not alliances, that Britain would ensure the survival of her world empire.[30]

III

The expansion of the world's navies and the rapid growth in the demand for guns and ships was soon felt at Elswick. Following the amalgamation with Mitchell in 1882, which integrated the company forward into ship building, Elswick continually expanded through internal growth and amalgamations during the prewar era. The amalgamation with Whitworth's in 1896 eliminated a strong competitor and prevented another firm like Vickers from absorbing it. The main asset was Whitworth's steel plant at Openshaw, near Manchester. Hamilton Rendel was against the amalgamation because he thought it would prompt Woolwich to expand its plant, but, as Armstrong told Stuart Rendel, he had 'far less fear of Government rivalry than of well-organized private enterprise'.[31] Elswick moved into extremely profitable armour plate production and developed ties with other companies to secure its supply of iron ore from Spain, coal from Wales and nickel (an alloy in armour) from Canada. By the end of the 1890s, with the single exception of engines, which it left to others to manufacture, Elswick produced complete warships of all sizes in its own facilities. Meanwhile, armour plate manufacturers were integrating forward into ship building and the already intense competition in the British ship-building industry became fierce.[32]

The need for economies of scale drove armaments and ship-building companies into agreements with each other. A system of production agreements and interlocking directorships rapidly evolved between Armstrong, Vickers (who moved from steel plate manufacturing into gun production), John Brown, Thomas Firth, Palmer and Beardmore. Elswick entered joint agreements to establish a number of dockyards and arsenals around the world: with Vickers in the Whitehead Torpedo Company in Fiume (Austria) in 1906; with Vickers, John Brown and some Spanish businessmen in the Spanish Sociedad Española de la Construcción Naval in 1908; with Vickers and Hokkaido Tanka Kissen in the Japan Steel Works in 1907; with Vickers and the Turkish government in the Turkish Dock Company in 1913; and two other torpedo factories with Vickers: the Société Française de Torpilles Whitehead in 1913 and the Società Anonima Italiana Whitehead in 1914.[33] Elswick, however, retained complete control of its major plants at Pozzuoli, rejecting an offer from Terni to buy a share.[34] In 1903, Elswick entered another partnership with Albini, the industrialist Ferdinando Perrone and the Bombrini family and acquired control of the Genovese naval yard and engineering plant at Ansaldo. The new company was called Ansaldo-Armstrong.[35]

Elswick was a moving force in the creation of an integrated worldwide armaments industry. Armour plate production came under international control through agreements among the private producers, the most infamous being the Harvey United Steel Company, a trust set up in 1901 by English, French, German and Italian armourers to 'amalgamate or control four other companies holding rights of the Harvey patents for treating steel'.[36] The Harvey Company was later held up as the greatest example of conspiracy by the world's armament kings.[37] John Falkner of Elswick calculated that, by controlling the world's armour plate supply, the 'Convention' maintained the average price at £115 per ton instead of the £40 per ton to which it would have fallen under a competitive system. After 1903, Elswick's Armour Plate Department at its Openshaw mill earned more profit than the Ordnance Department.[38]

The armaments kings also developed means to finance the purchase of armaments. The Société D'Etudes, formed by Armstrong, Vickers, the French Schneider company and a French investment group, Bardac & Company, provided loans for countries with little ready cash to buy weapons. The Société had a million francs of capital and made its first loan to China.[39]

Britain and Germany had always allowed Armstrong and Krupp to sell arms abroad, but it was not until 1886 that the French government freed the Schneider firm to export arms. This of course increased the competitive pressures and diplomatic complexities of the international trade in arms. The French introduced the tied loan, by which states wishing to borrow money on the French Bourse were required to purchase armaments in France. This strategy was used to bolster the French presence in the eastern Mediterranean against the growing German influence in Turkey. The sales rivalry between Schneider and Krupp in the Balkans was thus especially fierce, involving as it did the diplomatic and strategic goals of those two states. Each state acted according to its own traditions and financial capacity. The German government could not compete with French capital, but provided loans for private firms to build warships. On the other hand, French loans to Italy in 1904–6 played a role in weakening Italy's commitment to the Triple Entente, and she rejected a French loan in 1908 because the Italian army adamantly refused to switch from Krupp to Schneider artillery. At the same time, the French government refused to allow Austria and Hungary to raise money on the Paris stock market despite the diplomatic possibility of enticing Austria away from Germany and despite the loans being tied to the purchase of French armaments. Of course pulling Austria from Germany towards France was unlikely, and the refusal may well have had the objective of forcing cash-short Germany to lend Austria money (which the German Foreign Ministry successfully insisted it do) at the cost of postponing a loan to Turkey and Greece. Similarly,

Austria imposed tariffs to prevent Serbia from switching its arms purchases from Skoda to Schneider and from establishing closer economic ties with Bulgaria. The diplomacy of international armaments sales was indeed complex, and played its part in exacerbating the hostilities of the prewar years.[40]

The United States entered the global arms market in 1910, when the Fore River Company, backed by a diplomatic campaign waged by the United States State Department, beat its European competitors to win a contract to build two Dreadnought battleships for Argentina.[41] The American attempt to sell arms to Turkey failed completely because the United States had no diplomatic cards to play.[42] Britain continued to favour no European diplomatic 'entanglements' and was less interested in supplying allies than in maintaining British naval superiority. Nevertheless, the government lent its diplomatic support to Armstrong and Vickers in modernising the Turkish navy.

IV

The sale of armaments abroad was regulated by the government only during wartime. Through the Foreign Enlistment Act (1870) and the Customs and Inland Revenue Act, the government prohibited arms shipments to belligerents.[43] As an international crisis developed in which Britain had an interest, the government kept an eye on armaments exports. On 26 April 1877, two days after Russia declared war on Turkey, British Foreign Secretary Lord Derby asked the Treasury to instruct its customs officers 'to report at once and from time to time whether any vessels of War are being constructed at Ports within their districts' so that he could decide whether or not the Foreign Enlistment Act should be applied.[44] The Treasury had informed the secretary of state for war about an Elswick gunboat shipment two days earlier.[45] The Proclamation of Neutrality, which put the Foreign Enlistment Act into play, passed Cabinet on 28 April. Within days, Armstrong's solicitors were arguing that he be allowed to deliver two ironclads to Turkey on the grounds that the contract had been signed before the war began. The two ships cost £76 350 and Armstrong claimed he would lose 5 per cent of that if delivery was a month late and 2½ per cent for each succeeding month. Armstrong's solicitors explained to Derby that the company 'have been assured and believe that neither of the Corvettes in question is in fact intended or likely to be employed in the Military or Naval Service of either belligerent'.[46]

Elswick's worldwide foreign sales made it inevitable that a working relationship with the government would develop and Armstrong cultivated a cooperative relationship between Elswick and the government by keeping them informed of its sales abroad during moments of crisis. In 1881, for example, amidst rumours of war between China and Japan, Sir William wrote to the secretary of state for war and informed him, 'in accordance with our promise', that Elswick was about to deliver two gunboats to China. Granville was satisfied that war was not imminent and therefore the sale could be completed.[47] At the height of the Boxer uprising in 1900, Elswick voluntarily forwarded a list of Krupp and Armstrong guns sold to China between 1894 and 1900.[48] Other kinds of cooperation with the government were developed. For example, the company provided the government with important information about the effects of Japanese shells on the Russian fleet at Port Arthur in the attack of 1904 which began the Russo-Japanese War. The Japanese fleet carried mostly Elswick guns and a representative of the firm remarked in his report to the company that 'the Admiralty will I presume get from you any information you have'.[49] Elswick did pass on the information to the Admiralty.[50]

If British interests were not at risk, the government allowed the export of armaments whatever the consequences for others. In 1873, the viceroy of the Quang Province requested the British government stop the sale of arms to Chinese subjects without the authorisation of the Chinese government. But Chinese protests carried no weight and Granville decided that the sales did not threaten China's stability. A ban on the export of arms in this case 'would be attended with considerable difficulty and might create an inconvenient precedent in the future'.[51]

Some Elswick warships destined for other countries were purchased by the British government either to prevent the sale or because the Admiralty wanted the ship itself immediately, or both. Italy was deprived of a ship in this manner in 1888. To placate the Italians, Noble offered to build them another ship within 12 months and two within 15 months.[52] In 1904, the British government bought two Chilean battleships before they left England to ensure that they did not end up in Europe.[53] Europe in this case probably meant Turkey, which suffered a similar confiscation of a battleship in 1912.

Diplomatic support became increasingly important to the sale of armaments abroad after 1880. Traditional British foreign economic policy had been dedicated to the development of free trade, but did not concern itself with particular problems of individual companies.[54] A few individual diplomats such as the British minister in China, Sir Thomas Wade, took an active interest in selling Elswick armaments to China,[55] but generally British ambassadors, many of whom were upper-level aristocrats, considered it demeaning to their status to intervene directly on behalf of a specific firm. Other nations, particularly Germany and the

United States, did not hesitate to apply diplomatic pressure to get orders for their nationals and complaints from British businessmen multiplied over the lack of similar support from their government. Armstrong's agent in Turkey was particularly embittered by the strong diplomatic advantage enjoyed by Krupp:

> it requires a powerful combination to counteract anything emanating from His Highness Prince *Bismark* [*sic*] [and] any German subject who has business with or claim upon the Turkish government gets the most powerful support from his ambassador here. We get absolutely *None*.[56]

Similar complaints were expressed by other Elswick agents as Elswick's share of declining foreign markets was threatened by the direct German diplomatic pressure applied on behalf of Krupp.[57] Noble wanted Rendel or Sudeley (one of Elswick's aristocratic board members in Parliament) to do something, but he was not sure what.[58]

The policy was changed when the government realised that it was in their interests to promote exports of armaments. In 1885, Salisbury instructed British diplomats to become more directly involved in the problems of British foreign trade and commerce.[59] There is some evidence that Elswick had focused Salisbury's attention on armaments exports. In 1885, the company wrote to him and asked that the Foreign Office support its tender for some guns for the Romanian government.[60] Salisbury agreed and the first direct political support given to Elswick by the British government as a matter of policy was carried out by the prime minister himself.[61] Although many British diplomats resisted this new policy – the British minister in Chile reported in 1890 that the policy of non-interference remained in force[62] – the trend was towards greater diplomatic involvement in selling British arms abroad.

Government support of Elswick's foreign sales increased rapidly after 1900. This is particularly evident in Turkey. World demand for armaments had been falling since the late 1890s,[63] and Turkey became the only expanding market among the smaller powers, attracting arms merchants from all over Europe and the United States.[64] Despite the financial risks (Turkey still had difficulty paying for its weapons) sales were vigorously pursued. Success depended upon strong diplomatic support, and parliamentary declarations were made on behalf of Elswick.[65] In 1902, Lansdowne, the British foreign secretary, instructed Sir Nicholas O'Connor, the British ambassador in Constantinople, to assist Elswick in obtaining overdue payments. But O'Connor was of the old school and did not think the Foreign Office existed to serve commercial interests. 'As a rule', said John Falkner, the senior Elswick executive in charge of sales to Turkey, 'he shows himself difficult and contemptuous of all things commercial'.[66] O'Connor failed to see why

Elswick bothered with bankrupt Turkey anyway. He reported that 'the English firm however persisted in pressing for it, and with the assistance of the Embassy they obtained their desire'. Elswick received a £98 791 payment from Turkey.[67]

But the ambassador's help came at a price. O'Connor, probably with the agreement of the Foreign Office, took the opportunity to collect the overdue indemnity from Turkey for property lost by British businesses during the Armenian massacres of 1895–6. For O'Connor's diplomatic support, Elswick paid the Embassy £63 000, which was distributed among the claimants.[68] Once this was settled, O'Connor lost interest in Elswick. However, he was continually pushed into helping by both Conservative and Liberal foreign secretaries. In 1907, Elswick again complained to the Foreign Office about the lack of diplomatic support from O'Connor,[69] and instructions were sent out again to the ambassador to support Elswick fully in its efforts to sell warships to Turkey.[70]

Turkey presented a particularly sensitive diplomatic problem. German influence had increased in the 1890s and Russian sensibilities had also to be taken into consideration. After weeks of negotiating in the blazing summer heat of Constantinople, Falkner's efforts were upset 'by a *strong* representation that the purchase of a battleship by Turkey would be a "menace" to Russia'.[71]

British armaments and diplomacy became more tightly bound to each other during the Edwardian era, a combination which received support from the press. In 1911, even the radically anti-militarist *Economist* agreed that the Foreign Office should 'place their diplomats at the service of our own manufactories in order to secure a fair share of lucrative contracts'.[72] Britain's strategic interest in maintaining a large capacity to produce armaments coincided with Elswick's monetary interest in developing strong foreign markets. Elswick's executives and Britain's foreign secretaries were thus often found huddled together for their own reasons. As Falkner once quipped to Rendel, 'I am dining with Sir Edward Grey tonight to talk Turkey'.[73]

V

Two opposing views have been developed about the nature of the relationship between British armaments companies and the British state in the era of naval armament races which preceded the First World War. The first emerged just prior to and during the war and was revived in the 1930s. Private armaments companies, the argument ran, formed a vast international conspiracy which, through price fixing and the fomenting of wars and war scares, virtually forced governments to buy unnecessary quantities of weapons of unprecedented destructive power. The sharing of international markets, price and production agreements, and the willingness to sell to any and all sides in wars were held up as proof that the large armaments companies put profits above competition and even patriotism. The conspiracy took on greater proportions when the management and shareholder lists of these companies revealed the names of admirals, generals and politicians. Although their argument suffered from extrapolation, exaggeration and near-hysterical rhetoric, the authors made some telling points about the economic power and political influence of armaments industrialists.[74]

This 'merchants of death' thesis was not answered until the 1960s, when two business historians developed arguments which emphasised the power of the government as the only consumer of armaments. Clive Trebilcock, author of a very sympathetic work on Vickers, argued that the British government remained firmly in control of armaments procurements, that indeed it forced the private firms to comply with its technological designs and prices, and that it set production quotas and maintained healthy competition.[75] Robin Higham concluded that the most salient feature of the relationship was 'the omnipresent nature of the government as customer, paymaster, and regulator' which exercised 'virtually monopolistic powers, forcing companies to tailor themselves to its demands and the whims of its servants'.[76] At the same time, however, Higham conceded that the government's attempt to drive prices down by creating new armaments companies and maintaining its own factories had, in many cases, the opposite effect. Government factories remained technologically unprogressive and inefficient and the new companies required an enormous start-up capital investment and were managed by men with little business experience. These factors pushed prices up, which only benefited the few older companies.[77]

Arthur Marder struggled with the nature of this relationship and concluded that, although private arms manufacturers had great economic and political power, the great naval expansion which began in the 1890s was driven, not by the profit seekers but by the wave of imperialism that swept across Europe and Britain in the 1880s and 1890s. Imperial expansion required a larger navy and 'the munitions makers were only supplying a demand ... they were in themselves only the instruments of policy'.[78]

Whether they emphasise the role of private arms makers in pursuit of profit or governments in pursuit of power, all investigators agree that a close relationship of some kind developed between the private manufacturers of guns and ships and the British government during the 1880s and 1890s.[79] The influence of the private companies on British naval policy and expansion was based on two things. First was the desire of governments to expand the productive capacity for armaments in the country without taking on the high capital costs and financial risks involved. Lord Northbrook saw more than economic and political benefits from increasing the work done in private yards. British security, he said, was increased by allowing private ship builders to gain experience in building warships which it would be necessary to call into immediate action when war broke out. For this reason the Admiralty had recommended that private yards build new ships and the Royal Dockyards take on all the work of repairing ships.[80] By making the private yards part of the 'Naval Establishments', Northbrook provided arms manufacturers with the argument that they contributed to Britain's naval supremacy and the security of the Empire and therefore it was the duty of the government to maintain the demand for privately-produced armaments at home and, since the home market was too small to satisfy the economies of scale and bring an adequate return on investment, to promote the sale of British arms abroad.

The second source of influence of private firms lay in the fact that they had the experience and knowledge with which new weapons were developed. The Foreign Office and the Admiralty, together with the prime minister, decided British foreign and naval policy, but foreign policy objectives were achieved and naval strategies carried out by the evolving naval technologies of private armaments companies.[81] With the exception of hull design, private firms made virtually all the significant technological innovations upon which the British navy was built. New guns, gun mountings, armour, shells, torpedoes, engines and boilers all came from private innovators.[82]

Armstrong advocated, as he had always done, that the War Office and the Admiralty 'act in concert' with Elswick on designs for new guns. He was less prone to fight than Rendel:

> If the War Office and Admiralty do not choose to insist upon this joint action [he wrote to Noble] we can do nothing more than manufacture to specification and we must be content to do that if we are only to be treated as contractors.[83]

Armstrong envisioned a working relationship with the War Office and the Admiralty in which there was a clear division of labour between research, development and production. He explained to a government inquiry that, in his opinion,

> the proper function of a Government establishment is to establish patterns, and to carry on experiments continually. The Government should manufacture to such an extent as to establish a certain standard of quality and good manufacture, but the main production ought to rest with the outside manufacturers.[84]

In the end, the technological development of British armaments fell almost entirely to Elswick and other private firms.

Elswick was in the forefront of a process by which privately developed innovations often created confusion among those responsible for the navy. The most dramatic examples (next to the effect of the Armstrong gun on naval construction) were the development at Elswick of the heavily-armed, unarmoured *Esmeralda* cruisers and the development of the Whithead torpedo and the fast torpedo boat. In the 1880s, many experts, especially in France, became convinced that torpedo boats and cruisers had rendered battleships obsolete. The French stopped making battleships in 1881 and accepted the naval strategy, advocated by the *Jeune Ecole* faction in the French navy, of the *guerre de course* (war on commerce), by torpedo boats and fast, heavily-armed cruisers. The advocates of this new kind of naval warfare reached the height of their influence in 1886, when the *Jeune Ecole* leader, Admiral Hyacinthe Aube, was appointed minister of marine. In England, the old fear of invasion was replaced by the new fear of starvation, their food supply from the outside world cut off by French torpedo boats (and, after 1900, submarines) lying in wait for merchant ships to cross their paths.[85] The Admiralty was thrown into confusion over the correct response to the new threat from heavily-armed cruisers, quick-firing guns and torpedo boats. The First Lord of the Admiralty admitted in 1884 that, if the Admiralty was granted three or four million pounds, they would have difficulty deciding how to spend it.

Technological innovation had called into question all the old strategic and tactical certainties from which one determined what kinds of ships the Royal Navy needed. 'It was,' said Northbrook, 'very difficult to say what a first-class ship was.'[86] A debate broke out between those, led by Admiral Sir Cooper Key and Sir William Armstrong, who wanted Britain to increase the number of cruisers and develop a naval strategy to protect British international trade and commerce, and the majority inside the Admiralty, and many naval writers, like Admiral Sir Arthur Hood, Admiral Philip Colomb and Sir Charles Dilke, who insisted that Britain expand her battleship fleets and maintain her traditional role as the world's greatest sea power. The question remained under discussion throughout the prewar era but, following the failure of the *Jeune Ecole* to demonstrate the efficacy of torpedo boats in the French manoeuvres of 1887, the command-of-the-sea strategy prevailed.[87] The enormous success of Alfred T. Mahan's naval histories in the early 1890s, which advocated battleship warfare and command-of-the-sea strategy, entrenched the idea that the era of the great battleship was not drawing to a close but, indeed, had only begun.[88]

Armstrong strongly advocated a protection-of-trade naval strategy.

His position reflected his long-held belief in the primacy of the gun over armour and the fact that Elswick was the leading manufacturer and innovator of warships and guns to combat torpedo boats and battleships. The *Esmeralda* cruisers were faster than battleships and carried guns which were effective against the great capital ships and also the new quick-firing guns to be used against torpedo boats. He found it surprising that 'so much apathy should be displayed regarding the unprotected state of our commerce', and that, in the midst of the naval scare of 1884, 'the cry is chiefly for more ironclads, which are of little or no use for the protection of commerce'. Britain would 'actually starve' without imported food but she had too few swift cruisers in relation to the number of her merchant ships and the area of ocean to patrol. Armstrong dismissed the traditional tactic of arming merchant and passenger ships in an emergency, as 'they would simply supply victims for Esmeraldas'. Fast cruisers were much cheaper than battleships and if a large number were deployed against an ironclad they would prevail. Speaking at the annual meeting of the Armstrong, Mitchell Company in 1884, Armstrong concluded that 'these observations are pertinent to the occasion, because the success of the Esmeralda, and a more appreciative view of ships of the cruiser class, are obviously connected with the prospects of this company'.[89]

At every opportunity, Armstrong repeated his argument that the Royal Navy had not one ship in its fleet that could either evade or attack one of his cruisers.[90] In his first (and only) speech in the House of Lords, Armstrong stated again his belief that guns would always prove themselves against armour and consequently the addition of more cruisers was 'a much more important object than adding further to our fleet of ironclads'. Heavily-armed cruisers threatened the British Empire, he said, and party interests must not be allowed to impinge upon strategic considerations if the Empire and British prestige were to be maintained in the new age. Armstrong, mindful of where he was, trusted that his call for a protection-of-commerce naval strategy would not be seen as self-interested. 'But in any case,' he said, reflecting another of his long-held views, 'it is facts and arguments, and not motives that must determine the issue.'[91] His recognition that facts did not speak for themselves was something he learned from Stuart Rendel during the competition with Whitworth in the early 1860s over what constitutes the 'best' shell to sink ironclads. He never forgot that lesson. One could say that, before 1864, Armstrong was a technological determinist, but the competition with Whitworth gave him a social constructionist appreciation of engineering and enterprise. Nevertheless, the statement indicates that Armstrong retained his modernist view that objective engineering judgments were possible.

But it was the age of battleships and command-of-the-sea and Armstrong did not let his preference for cruisers and commerce protection prevent Elswick from becoming an important part of it. Elswick was soon building battleships for Britain and the lesser nations of the world. The Japanese navy bought a number of Elswick battleships; Argentina, Brazil and Chile bought several each and Norway and Romania at least one each.[92]

In the 1890s, the battleship emerged as a symbol of a nation's technological sophistication and industrial capacity as well as a sign of its place in the hierarchy of nations. The great powers built their own ships and competed amongst themselves to maintain or improve their positions; the lesser nations purchased their power and prestige from the industrial giants and conducted their own rivalries. Elswick was a crucial part of this worldwide competition for naval greatness, the place at which many of those symbols were developed and constructed. Indeed, the true symbol of British naval power was not the great battleships on the seas but the great armaments factories at Elswick.

Notes

1. Paul Kennedy, *The Rise and fall of British Naval Mastery* (London: Fontana, 1991), p. 205.
2. A.J.P. Taylor, *The Struggle for Mastery in Europe 1848–1918* (Oxford: Oxford University Press, 1971), p. 215. As Taylor put it: 'The Russians wanted the other powers to recognize their right to keep warships [in the Black Sea], not actually to have them', and adds the counter-factual speculation that, if the Russians had had a fleet in the Black Sea in 1877, 'the war [with Turkey] would soon have been over, with immeasurable consequences for the history of Europe'.
3. John Colomb, *The Protection of Our Commerce and the Distribution of Our Naval Forces Considered* (1868).
4. Arthur Marder extends the period of neglect back to 1815 with his witty comment that 'from the conclusion of the Napoleonic wars until late in the century the subject of naval defence was almost foreign to English thought'. See his *The Anatomy of British Sea Power: A History of British Naval Policy in the Pre-Dreadnought Era, 1880-1905* (New York: Alfred Knopt, 1940), p. 44.
5. Oscar Parkes, *British Battleships "Warrior" 1860 to "Vanguard" 1950: A History of Design, Construction and Armaments* (London: Seeley Society, 1966).
6. Theodore Ropp, *The Development of a Modern Navy: French Naval Policy 1871–1904* (Annapolis, Md.: Naval Institute Press, 1987), p. 79.
7. Kenneth Warren, *Armstrongs of Elswick: Growth in Engineering and Armaments to the Merger with Vickers* (London: Macmillan, 1989), chs 10, 11, 16.
8. Ropp, *Modern Navy*, p. 79.

9. Paul Kennedy, *The Rise of the Anglo-German Antagonism 1860–1914* (London: George Allen & Unwin, 1980), pp. 252, 254, 370.
10. Harvey Blumenthal, 'W.T. Stead's Role in Shaping Official Policy: The Navy Campaign of 1884', PhD thesis, George Washington University, 1984, pp. 54–7.
11. O.J. Hale, *Publicity and Diplomacy* (Gloucester, Mass., 1964), p. 13. Stead lost influence when he went against public opinion in opposing the Boer War. He missed the greatest truth story about the *Titanic*, going down with that ship.
12. Steven R.B. Smith, 'Public Opinion, the Navy and the City of London: The Drive for British Naval Expansion in the Late Nineteenth Century', *War and Society* 9 (May 1991), p. 37. Smith identifies some other independent-minded officers who gathered around Phipps-Hornby as Captain James Bruce, Charles C. Penrose Fitzgerald and probably Eardley-Wilmot. P.H. Colomb and Lord Charles Beresford were also leading individuals in the naval reform movement. Smith sees the real influence for change coming from an alliance between the City and the armed forces, with the latter providing the greatest political pressure on Gladstone's government.
13. Blumenthal, 'Stead', p. 65.
14. Ibid., p. 116.
15. Blumenthal, 'Stead', pp. 66–7.
16. *The Times*, 2 October 1884, p. 2.
17. Smith, 'Public Opinion', p. 30.
18. Hamilton's diary, quoted by Smith, 'Public Opinion', p. 32.
19. 'The State of the Navy', *The Times*, 17 April 1885, p. 10.
20. Smith, 'Public Opinion', pp. 38–9.
21. *Hansard*, 16 March 1885, col. 1279.
22. India and China accounted for nearly a quarter of British exports; China and Japan imported more from Britain than from the rest of the world combined. This trade went through Suez. See M.C. Morgan, *Foreign Affairs 1886–1914* (London: Collins, 1973), pp. 13–15.
23. Marder, *Anatomy of British Sea Power*, pp. 105–111, 126–32; Paul Kennedy, *Strategy and Diplomacy 1870–1945* (London: Fontana, 1983), pp. 98–9; William L. Langer: *The Diplomacy of Imperialism, 1890–1902* 2nd edn (New York: Alfred A. Knopf, 1956), pp. 420–23; Lady Gwendolen Cecil, *Life of Robert Marquis of Salisbury*, vol. 4 (London: Hodder and Stoughton, 1932), pp. 183–93.
24. Marder, *Anatomy of British Sea Power*, ch. 4.
25. Marder, ibid., ch. 10.
26. The literature on international relations between 1880 and 1914 is very large. Good introductions are Joachim Remak, *The Origins of World War I, 1871–1914* (New York: Holt, Rinehart and Winston, Inc., 1967); C.J. Bartlett, *The Global Conflict: The International Rivalry of the Great Powers, 1880–1970* (London: Longman, 1984), chs 1–3; Richard Langhorne, *The Collapse of the Concert of Europe: International Politics, 1890–1914* (London: The Macmillan Press Ltd., 1981); and Taylor, *Struggle for Mastery in Europe*. More detailed studies are found in two books by William L. Langer: *The Diplomacy of Imperialism, 1890–1902* and *The Franco-Russian Alliance, 1890–1894* (Cambridge, Mass.: Harvard University Press, 1929). Langer's books include discussions of the naval imbalances and strategic considerations of the European rivalry. See,

for example, *The Franco-Russian Alliance*, pp. 336–49, 355–66, 390–91. Naval strategic affairs are also examined by Marder in *The Anatomy of British Sea Power* and *From The Dreadnought To Scapa Flow: The Royal Navy in the Fisher Era, 1904–1919*, vol. I, *The Road to War, 1904–1914* (London: Oxford University Press, 1961). Three good books on the naval arms race between Britain and Germany are Peter Padfield, *The Great Naval Race: The Anglo-German Naval Rivalry, 1900–1914* (London: Hart-Davis, MacGibbon, 1974); Jonathan Steinberg, *Yesterday's Deterrent: Tirpitz and the Birth of the German Battle Fleet* (New York: The Macmillan Company, 1965); and E.L. Woodward, *Great Britain and the German Navy* (London: Frank Cass and Co. Ltd., 1964). For a broader view, see Paul Kennedy, *The Rise of Anglo-German Antagonism, 1860–1914*.

27. *Hansard*, 2 Dec. 1884, col. 410.
28. Lord George Hamilton, *Parliamentary Reflections 1886–1906* (London: 1922), pp. 220–21, quoted by William H. McNeill, *The Pursuit of Power: Technology, Armed Force, and Society Since A.D. 1000* (Chicago: University of Chicago Press, 1982), pp. 275–6. Hamilton considered this sentiment of the working classes towards the navy as independent from the fact that most of them escaped the taxation which paid for it. See also the comments by one M.P. that the two million new voters 'would not protest against any reasonable expenditure of money for the preservation of the honour of the Empire' (*Hansard*, 16 March 1885, col. 1285).
29. McNeill, *Pursuit of Power*, pp. 269–78.
30. Marder, *Anatomy of British Sea Power*, ch. 15. An attempt to generate an invasion scare in 1900 fell flat. See Marder, ibid., ch. 28.
31. Armstrong to Rendel, 24 Dec. 1896, T&W, 31/3684.
32. Sidney Pollard and Paul Robertson, *The British Shipbuilding Industry: 1870–1914* (Cambridge, Mass.: Harvard University Press, 1979), p. 270, fn. 28.
33. R.P.T. Davenport-Hines, 'Vickers As a Multinational Before 1945', in Geoffrey Jones (ed.), *British Multinationals: Origins, Management and Performance* (London: Gower, 1986), pp. 46–7; J.W. Walton, *The War Trust Exposed*. Disputes over management positions are discussed in E.D. Boyle to Elswick, 11 Jan. 1911, T&W, 31/7005. See also 31/7772, /7786, 7792, /7803, /7804 and /7806.
34. Terni turned to Vickers and the Vickers-Terni Company was formed in 1906. See Luciano Segreto, 'More Trouble than Profit: Vickers' Investments in Italy, 1906–39', *Business History* 27 (1985), pp. 317–18.
35. Ibid., p. 318.
36. T&W, 31/7033 (16 July 1901). Albert Vickers was chairman and John Falkner represented Elswick.
37. H.R. Murray, *Krupp's and the International Armaments Ring*, pp. 150–61. Pollard and Robertson, *British Shipbuilding*, p. 98, conclude that these agreements 'make it permissible to speak of a single international warship and armaments combine'.
38. R.J. Irving, 'New Industries for Old? Some Investment Decisions of Sir W.G. Armstrong, Whitworth and Co. Ltd., 1900–1914', (*Business History*, 17, July 1975, p. 173, fn. 54).
39. T&W, 31/7035 (n.d.).

40. James Joll, *The Origins of the First World War* (London: Longman, 1984), pp. 126–38.
41. *New York Times*, 23 Jan. 1910, Section C, p. 4, 26 Nov. 1911, Section C, p. 4.
42. 'Our Shipbuilders Cut Out', *New York Times*, 18 March 1910, p. 4. As the newspaper observed, contracts awarded without competition were not unusual, but, with the modernisation of Turkey, 'much hope had been felt of much closer relations between the United States and the Ottoman Empire'.
43. *Parl. Papers* II (1878–9), p. 693.
44. Derby to Treasury, 26 April 1877, F.O. 78/2665. Spying for the Foreign Office was still practised by Customs 10 years later. See London Customs House to F.O., 19 Dec. 1888, F.O. 46/384, f.389.
45. Ibid., f.253.
46. Radcliffe, Cator and Martineau to Derby, 3 May 1877, F.O. 78/2664. Thorneycroft, the torpedo boat manufacturer, was also anxious about his contract with Turkey. See F.O. 78/2665. This file contains several references to arms shipments to Turkey, including the Proclamation of Neutrality approved by the Cabinet on 28 April 1877.
47. Armstrong to Earl Granville, 20 April 1881, F.O. 17/870, f.255, 258.
48. Noble to Francis Bertie (Assistant Undersecretary of State for Foreign Affairs), 3 Nov. 1900, F.O. 17/1449, f.43.
49. Edward Boyle to Armstrong & Co., 31 May 1905, Dey/9. Boyle hid in Yokohama to avoid others in the armaments community who could pressure him for the information.
50. Hugh Lyon, 'The Relations Between the Admiralty and Private Industry in the Development of Warships', in Brian Ranft (ed.), *Technical Change and British Naval Policy, 1860–1939* (London: Hodder and Stoughton, 1977), p. 57. Elswick had shared information with the War Office as far back as 1866, when Stuart Rendel informed them about Russian experiments on rifled guns. See Major General St. George (Director of Ordnance) to Rendel, 10 July 1866, T&W, 31/715.
51. F.O. 17/692.
52. Noble to Rendel, 15 June 1888, T&W, 31/3048.
53. *Hansard*, 23 Feb. 1904, cols 742–3, 751.
54. See Palmerston's Circular to British diplomats, 15 Jan. 1848, *Parl. Papers* LVI (1849), pp. 1–4, reprinted in Kenneth Bourne, *The Foreign Policy of Victorian England 1830–1902* (Oxford: Oxford University Press, 1970), pp. 288–9.
55. S.W. Bridgford to Elswick, 1 Nov. 1881, T&W, 31/4187.
56. A.H. Leak to Elswick, 20 Oct. 1888, T&W, 31/3077. See also a similar report from A.H. Scaife, 6 Nov. 1888, T&W, 31/3082.
57. Falkner to Rendel, 5 Oct. 1888, T&W, 31/3071.
58. Noble to Rendel, 13 Nov. 1888, T&W, 31/3088.
59. D.C.M. Platt, *Finance, Trade and Politics: British Foreign Policy 1815–1914* (Oxford: Clarendon Press, 1968), pp. 304–5.
60. Noble to Rendel, 12 Sept. 1885, T&W, 31/2647.
61. Marvin Swartz, *The Politics of British Foreign Policy in the Era of Disraeli and Gladstone* (London: Macmillan, 1985), p. 211, fn. 21.
62. Captain Temple to British Embassy in Paris, 7 July 1890, F.O. 27/3016.
63. Noble to Rendel, 28 July 1901, T&W, 31/3600.

64. Falkner to Rendel, 10 July 1904, T&W, 31/7051.
65. *Hansard*, 21 June 1900, col. 627.
66. John Falkner to Rendel, 10 July 1904, T&W, 31/7051.
67. Letters from O'Connor to Lansdowne, 24 May and 22 Oct. 1902, F.O. 78/519/243 and 78/5193/470.
68. Some of the claimants are listed in Edward Blech (consul at Constantinople) to O'Connor, 29 April 1902, F.O. 78/519.
69. F.O. 371/343/15180.
70. For the correspondence between Sir Edward Grey, O'Connor, the Admiralty and Elswick see F.O. 371/354. For the role of diplomacy in Elswick's efforts to sell commercial steamships to Turkey in 1907 for use on the Tigris, see F.O. 371/343.
71. Falkner to Rendel, 4 Aug. 1904, T&W, 31/7053.
72. *The Economist*, 12 Aug. 1911, p. 324.
73. Falkner to Rendel, 21 June 1910, T&W, 31/7089.
74. The titles of the literature are suggestive: *The War Trust Exposed* (1913), *The War Traders: An Exposure* (1913), *Krupp's and the International Armaments Ring* (1915), *Iron, Blood and Profits: An Exposure of the World-Wide Munitions Racket* (1934), *Merchants of Death: A Study of the International Armaments Industry* (1934), *The Private Manufacture of Armaments* (1936).
75. Clive Trebilcock, 'A "special relationship" – Government, Rearmament, and the Cordite Firms', *Economic History Review* 19 (1966), pp. 364–79; 'Legends of the British Armament Industry 1890–1914: A Revision', *Journal of Contemporary History* 5 (1970), pp. 3–19; 'The British Armaments Industry 1890–1914: False Legend or True Utility', in Geoffrey Best and Andrew Wheatcroft (eds), *War, Economy and the Military Mind*, (London: Croom Helm, 1976), pp. 89–107; *The Vickers Brothers: Armaments and Enterprise, 1854–1914* (London: Europa, 1977).
76. Robin Higham, 'Government, Companies, and National Defence: British Aeronautical Experience, 1918–1945 as the Basis for a Broad Hypothesis', *Business History Review* 39 (1965), pp. 323–47.
77. Ibid., pp. 327–8.
78. Marder, *Anatomy of British Sea Power*, p. 42. Marder had published these conclusions two years earlier in 'The English Armament Industry and Navalism in the Nineties', *Pacific Historical Review* 7 (1938), pp. 252–3.
79. A judicious analysis of the interaction between private arms manufacturers and the British state between 1884 and 1914 is provided by McNeill (*Pursuit of Power*, ch. 8).
80. *Hansard*, 2 Dec. 1884, cols 409–10.
81. Maurice Pearton, *The Knowledgeable State: Diplomacy, War and Technology Since 1830* (London: Burnett, 1982), p. 44.
82. Lyon, 'The Relations Between the Admiralty and Private Industry', pp. 37–64.
83. Armstrong to Noble, 3 March 1883, T&W, 31/3133.
84. Armstrong testimony to the Royal Commission on Warlike Stores, *Parl. Papers* XV (1887), q. 9510.
85. Alan Cowpe, 'The Royal Navy and the Whithead Torpedo', in Ranft (ed.), *Technical Change and British Naval Policy*, pp. 23–36. See the chilling prediction by Admiral Aube of the sinking of a merchant steamer, in Marder, *Anatomy of British Sea Power*, p. 87.

86. *Hansard*, 10 July 1884, cols 661–2. See also his comments on torpedo boats, ibid., 2 Dec. 1884, cols 408–9. The split within the Admiralty caused by torpedo boats is outlined by Shaw Lefevre in *Hansard*, 10 June 1886, cols 1396–7. See also Marder, *Anatomy of British Sea Power*, ch. 6.
87. Bryan Ranft, 'The Protection of British Seaborne Trade and the Development of Systematic Planning for War, 1860–1906', in Ranft (ed.), *Technical Change and British Naval Policy*, pp. 1–22.
88. Marder, *Anatomy of British Sea Power*, pp. 45–8. Admiral Philip Colomb's books on 'sea power' argued the same case as Mahan did for 'command-of-the-sea', but Mahan's writing style and breadth of vision brought him the fame and adulation. See Langer, *The Diplomacy of Imperialism 1890–1902*, pp. 418–20.
89. *The Times*, 2 Oct. 1884, p. 2.
90. Parkes, *British Battleships*, p. 284.
91. *Hansard*, 9 Feb. 1888, cols 14–23.
92. *Jane's All the World's Fighting Ships* (London: Jane's, 1898).

CHAPTER NINE

Entrepreneurs, managers and the British state

This study has emphasised how technological innovation, political entrepreneurship and the international context determined the growth rate and overall business success of Elswick under the leadership of Sir William Armstrong. It was argued that the external relations of the company were more important than its internal command structure and individual personalities. Although external relations with the British state remained the single most uncertain variable for the company given, the question of corporate governance increased in importance as new members entered the company from firms which Elswick acquired through the years and as the founding members of the firm retired and died. But if top management determines the fate of companies, then, in the case of Elswick – and the other armaments firms – success went to those who best managed relationships with the state. The division between the external and the internal approach to the company breaks down only in the sense that the primary objective of management was to establish and maintain access to and the confidence of state mangers. This was in the hands of the individual men who managed the company for better or worse throughout the prewar era of great navies, men who not only oversaw the operations of the company in the technical sense but developed close personal ties with British governments, or brought into the company others who brought access to the highest levels of the British state.

A total of 31 men managed Elswick between 1883 and 1914 on a board of directors which fluctuated between a high of 19 members in 1897 and a low of eight in 1911. It was a group diverse in social background, political beliefs and practical experience: there were lawyers, engineers, diplomats, Members of Parliament (both Houses), Gladstonian Liberals, Unionist Liberals, Conservatives of various hues, military officers, aristocrats and commoners. One could claim to be a scholar, a novelist and a poet. They belonged to 26 clubs which represented a variety of social, political, scientific and literary interests. Personal qualities also ranged over a wide spectrum: temperaments differed, as did talent, strength of purpose, ambition and the intellectual and physical capacity for work. But the two things they had in common and which outweighed their differences were their commitment to

Elswick and their collective access to the full range of Britain's social and political elite and, most importantly, to the government of the day, whether Liberal or Conservative (or, later, Unionist). These common denominators were of great importance to the success of Elswick which, in 1905, became Britain's eleventh largest company.[1]

The first board had 14 men. Led by Lord Armstrong until his death in 1900, this group dominated the company up to the First World War.[2] Engineering abilities were of course prime considerations for board membership. Henry Swan and Charles Mitchell came onto the board as part of the 1882 amalgamation and remained responsible for Elswick's commercial ship building. Hamilton Rendel was in charge of the Engine Works. Armament production was directed by Armstrong and Sir Andrew Noble. George Rendel joined the Admiralty as Civil Lord in 1882, returned to the company in 1885, and moved to Italy to manage the new Pozzuoli works with Admiral Augustus Albini, whose continuing importance for Italian orders was recognised with a seat on the board in 1888. Percy Westmacott had been associated with Armstrong since 1851 and had worked his way up from draughtsman to manager of the Elswick works in 1858 and partner in 1863. He retained responsibility for the hydraulic machinery department.

Armstrong was always on the lookout for inventors of important new technologies and he did not hesitate to bring them into the company. Sir William Siemens was given a seat on the board for his knowledge of steel. Of equal importance was the engineering talent of Josiah Vavasseur, who had been Blakely's partner in the London Ordnance Works where he designed the carriages for the Blakely gun. Just as Armstrong's hydraulic machinery helped sell Armstrong guns, Vavasseur's carriages had made Blakely's guns attractive to the Confederate Navy in 1861 and the Russian Government in 1865.[3] Vavasseur was an innovative engineer and carried on the business after Blakely's death in 1868. In September 1882, he won his first large contract from the British Admiralty who considered his carriages and mountings to be better than those of Armstrong.[4] However, the £40 000 order required Vavasseur to expand his plant. He did not have the capital to do so and Firth, his steel supplier, was not interested in an amalgamation.

The logical alternative was an agreement with Armstrong but, as Vavasseur's lawyer, Bernard Broomhead, recalled, 'they were not on friendly terms, and they could neither of them approach each other'.[5] Broomhead suggested that he approach Elswick on behalf of Vavasseur and determine if Elswick was interested in negotiating an agreement. Vavasseur agreed and through a mutual friend Broomhead arranged to meet Westmacott on the train between London and Doncaster.

Westmacott reported to the board in November and a few days later Broomhead met Noble (also on a train) and on 4 December Broomhead introduced Vavasseur to Noble. At the first shareholders' meeting of the Sir W.G. Armstrong, Mitchell & Company on 22 February 1883, an agreement was signed by which Vavasseur sold his business to Armstrong and joined the new company. Elswick thus obtained an extremely valuable piece of technology along with its inventor. Josiah Vavasseur became a director of one of England's largest and most progressive companies.[6] At Elswick he found the resources with which to continue his research and development and in 1887 he produced a revolutionary recoil system for the Armstrong quick-firing gun which used the recoil energy to return the gun automatically to its firing position. A new type of ship, the destroyer, was developed for this gun.[7]

Financial management of Elswick remained with William Cruddas, who had taken over that responsibility in 1861 from his father, George Cruddas, one of the founders of the crane company in 1846. William remained director of finances until 1911. He was an active Conservative: a member of the Primrose League and chairman of the Newcastle Conservative Association in 1884. Cruddas's political activities culminated in his victory (by 308 votes) over John Morley in the 1895 election. He remained an MP for Newcastle until 1900, but during those five years he made no speech and asked only one question (in 1898). However, Cruddas remained an important bridge between Elswick and the highest levels of the Conservative Party.

Stuart Rendel was not a member of the board of directors, but his position at Elswick was always special. Apart from Lord Armstrong himself, Rendel was the most important link to the British social and political elite. He entered the House of Commons as a Gladstonian Liberal in 1880 and helped Morley win his first seat in 1882 by showing him through the Elswick factories. Election results for the Newcastle constituency were determined by Elswick, the largest ward, and the outcome at Elswick was determined by its most important group of voters, the Armstrong workers. Rendel's politics were too radical for some on the board, including Sir William, whose dislike of them was 'needlessly stirred' by other board members.[8]

The involvement of Elswick managers in party politics enabled the firm to develop and maintain access to the British government whatever party was in power. Thus, when Sir Stafford Northcote visited Newcastle in 1884, it was Cruddas, not the Liberal (Unionist) Armstrong, who ushered that influential Tory to his social functions.[9]

Disputes over business strategy caused far greater disruption than any ideological differences. These emerged when the new public company was created. Stuart Rendel had opposed the way the new company was financed. Some partners had to surrender part of their interest to provide the public shares and Rendel considered himself the principal victim of this and that he had sacrificed greater future profits than the others.[10] He thought Mitchell's Low Walker shipyard a bad investment and the presence of Mitchell and Swan on the board exacerbated his feelings. Twelve years later, Armstrong was still trying to convince Rendel that, although they had made mistakes in the past, the Mitchell deal was not one of them. He told his old friend that no one agreed with him and that his colleagues did not like his persistence in the matter.[11] In the same letter Armstrong berated Rendel's brother Hamilton for his opposition to the imminent amalgamation with Whitworth's. Hamilton, he said, had been 'habitually opposed to development and Elswick would have been dwarfed if governed by him'. His seclusion, lack of enterprise, his prejudices and bad temper were not qualifications of a good manager.[12]

A man who was not on the board of directors but who deserves special attention for his engineering contributions to the firm is Sir William White. White was a brilliant naval architect who went from the Royal School of Naval Architecture in 1867 directly into the Admiralty where he worked for Sir Edward Reed and his successor, Sir Nathaniel Barnaby, and became chief constructor in 1881, a position he had held in all but name since 1878.[13]

The Argentine government brought White and Elswick together. In 1879, Argentine arranged with the Admiralty to allow White to design an armoured cruiser for them. Elswick won the armaments order for the ship and thus worked closely with White on the project. Armstrong recognised White's talent and White discovered that at Elswick he could design a variety of warships for the world market (and make much more money than he did at the Admiralty). He agreed with Armstrong that fast, heavily-armed cruisers had become essential parts of modern navies.[14] White was soon on his way north to design the new Elswick shipyard in which his warships would be built. He had, he said, 'sole responsibility of the management of that section of the business'.[15] At Elswick White designed warships for many foreign countries and supervised the two thousand men employed in the shipyard.[16]

Elswick's reputation for innovative warships reached new heights under White's direction. In 1885, the new First Lord of the Admiralty, Lord George Hamilton, pleaded with Armstrong to allow White to return to the Admiralty to carry out the Northbrook rearmament programme. Armstrong's consent was imperative[17] and White was released on the condition that Elswick could consult with him from time to time.[18] The government also agreed to replace White with another rising naval architect at the Admiralty, Philip Watts. Watts was given a seat on the board of directors in 1895 and returned to the Admiralty in

1902. He went back to Elswick in 1916.[19] Another talented naval architect, Tennyson D'Eyncourt, also moved with ease between the Admiralty and Elswick. Sir William Smith, after an Admiralty career as the director of naval construction and then superintendent of construction accounts and contract work, joined Elswick in 1913.[20]

Military officers found new careers at Elswick. As with Armstrong and so many others, Charles Younghusband's career was radically changed by the Crimean War. In 1854, after 13 years observing magnetic and meteorological phenomena in Canada, Younghusband found himself on the battlefield at Inkerman. He survived and moved up the British military bureaucracy, becoming the superintendent of small arms contracts, a member of the Ordnance Select Committee from 1863 to 1867, the superintendent of the Royal Gunpowder Factory between 1868 and 1875 and finally superintendent of the Royal Gun Factories from 1875 until 1880, when he retired at age 61.[21] Younghusband's thorough knowledge of armament technology, his long and extensive managerial experience and his close contacts inside the War Office were needed at Elswick and Armstrong brought Younghusband into the company as a founding director in 1883. He remained there until 1897.

The career of Major-General Sir Henry Alderson followed a similar pattern. After serving in the Crimean War he joined the Artillery Department at Shoeburyness and rose up the War Office bureaucracy: director of artillery from 1888 until 1891 and president of the Ordnance Committee between 1891 and 1896. He joined the Armstrong board in 1897.[22] Colonel Sir George Hadcock of the Royal Artillery moved from the Ordnance Committee to the board of directors at Elswick in 1916.[23]

Elswick men began to appear in Parliament. Armstrong had always had friends in Parliament, but after 1880 more direct links were established. Stuart Rendel was the first to be elected to the House of Commons. But Rendel did not spend his time in Parliament demanding that the government buy more armaments. He represented the Welsh constituency of Montgomeryshire and soon led the parliamentary branch of the movement for Welsh disestablishment. He was also largely responsible for the passage of a Welsh Education Act in 1889. The Celtic fringe had become important to the Liberal Party and Rendel, for example, was given the honour to move the Address in reply to the speech from the Throne in 1881.[24] Rendel and Gladstone developed a close personal relationship,[25] and Gladstone often recuperated at Rendel's estate at Cannes. One of Gladstone's final acts as prime minister in 1894 was to send Rendel to the House of Lords and a few years later Rendel carried Gladstone's coffin. The Rendel and Gladstone families became close too: in 1890, Gladstone's son, Henry, married Rendel's second daughter and, in 1902, Rendel personally provided Henry with a

£30 000 second mortgage.[26] In 1903, Henry joined the Elswick board of directors.

Rendel worked on behalf of China during the Sino-Japanese War of 1894–5. Through Robert Hart he became an intermediary between the British and Chinese governments during the attempts to stop the war.[27] But Rendel, a faithful Gladstonian, was barred from the government's inner circle and Rosebery decided that Britain's interests lay with Japan. However, Rendel enjoyed a close relationship with Campbell-Bannerman.[28]

Sir William Armstrong had no political ambition beyond securing the future of his company. He had been a Liberal all his life but had not taken an active role in party affairs and indeed had turned down requests from both Liberal and Conservative parties to run for Parliament. However, he was a committed Unionist and was persuaded in 1886 to let his name stand as a Liberal–Unionist in Newcastle. He lost to John Morley but was called to the House of Lords a year later by Salisbury. His only speech is more about the need for cruisers than about the need to keep Ireland.[29]

Edward F. Leveson-Gower joined the board in 1884. His attachment to Armstrong can be traced back to 1862, when, as a backbench Whig, he had argued against all patent laws.[30] Leveson-Gower had no experience in business or technology, but his political and social connections were useful attributes for the company. Of these he had plenty: son of the first Earl Granville, great-nephew of the fifth Duke of Devonshire, and Member of Parliament since 1846.[31] When he died in 1907 his son, George, wanted to take his place on the board. His business qualifications were no better than those of his father but, as Henry Gladstone reminded Rendel, George Leveson-Gower was a Liberal 'who might be useful perhaps in approaching officials'.[32] However, the company was in the midst of a factional struggle at the time (as we shall see shortly) and another friend of Stuart Rendel on the board was not possible.

Three aristocratic members of the first board of directors were Lord Thurlow, Lord Sudeley and Lord Magheramorne. Thurlow's lineage dated back to the sixteenth century and his long diplomatic, administrative and business experience complemented his aristocratic status.[33] Lord Sudeley had defended the Armstrong reputation in 1879 during the *Thunderer* crisis and was a strong advocate of the Admiralty's right to purchase guns on the open market (which meant Elswick).[34] Sudeley was an important Elswick link to the international powder industry. As part of an 1885 agreement under which the British government bought powder from the Chilworth Powder Company, an English subsidiary of a German powder group controlled by the Nobel Dynamite Trust,

Sudeley became chairman of the company.[35] In 1887, Chilworth began supplying Elswick with gunpowder. Sudeley resigned from Elswick to become the paymaster-general of the Forces.

Lord Magheramorne (Sir James Hogg) was a Conservative, 'but not opposed to progress',[36] and 'in favour of non-intervention'.[37] Hogg served in the army until 1859, when he began his political career which took him to the House of Lords in 1887. As chairman of the Metropolitan Board of Works, Hogg was involved in various schemes to develop London.[38]

The appointment of these aristocratic directors caused some resentment at Elswick. Sir James Hogg particularly offended Hamilton Rendel, who wanted to practise engineering and became frustrated with the increasing correspondence and other paper work associated with an expanding company. He complained to his brother that the outside directors did not realise 'what the work in the Ord. Dept. really is'. It angered him 'to think of that old Hogg sitting smugly and fancying that he is an important member of the company'.[39]

The Elswick aristocrats were not on the board of directors to attract investors. Armstrong's company had plenty of prestige and had no difficulty raising capital. The company had enough engineering talent, but it needed to promote its products at the political level. Sudeley, for example, spoke in the House of Lords on behalf a British naval strategy built around heavily-armed fast cruisers rather than the great battleships.[40] But the prime purpose of men like Sudeley and Thurlow was to advocate the use of British diplomacy on behalf of Elswick's foreign business.[41] One of the most diplomatically delicate armaments markets was the Balkans, where the interests of the great powers converged.

Romania is a good example. In 1878, the Treaty of Berlin had given Romania independence and an equal seat on the international Danube Commission.[42] Romania thus became a new customer for guns and warships to protect its sovereignty and carry out its part in patrolling the Danube. A fierce competition broke out between Armstrong and Krupp for this new market.[43] The British diplomatic position was crucial to Elswick's chances of obtaining its share of the Romanian market, but the British government supported Russian ambitions in Romania and the Romanians thus looked to Austria and Germany for diplomatic support.[44] In return, Bismarck demanded that the Romanians buy Krupp armaments.[45] Many within the Romanian government and military preferred Armstrong guns,[46] but they needed the British Foreign Office to issue a statement, however moderate, which supported Romania and could be used to counter the political pressure to buy from Krupp. Dyer wrote to Rendel and explained that, if Rendel

> could, yourself or through Lords Thurlow and Sudeley raise some political interest in the future position of the Riviera states, or even make a show of raising the question it could not fail to strengthen our hands very materially and place our relations with Rumania on a friendly footing: of course this is a delicate matter and one in which you will be far better able to decide on the line of action than we can here.[47]

It was a delicate matter indeed, apparently too delicate for open debate.[48] But no doubt Elswick's parliamentary representatives practised quiet diplomacy.

The private sector provided Elswick with innovative engineers like Siemens and Vavasseur, but the company looked to the British state for administrators and financial managers, and if a candidate was politically well-placed so much the better. Henry Gladstone's political credentials were obvious, and his close personal relationship with Stuart Rendel was an advantage, but his wide business experience and his work on various financial committees of the government made him particularly attractive to the firm. With Rendel's support, he became an outside director in 1903.

On 27 December 1900, Sir William Armstrong died and the firm entered a period of internal strife and factional struggles for power. Sir Andrew Noble, Armstrong's first recruit from the British military, who had worked at Elswick with Sir William for 40 years and had been vice-chairman since 1883, became chairman. He centralised power around himself and placed men who were loyal and obedient to him on the board of directors. In 1901–2, he made his son, Saxon, a managing director and Elswick's representative on the boards of the Mond Nickel Company and the Whitehead Torpedo Company.[49] Noble virtually forced the board to accept John Falkner as a managing director who, with Noble, took charge of the Ordnance Department.

Falkner was not the kind of person one envisions as a senior executive of a huge armaments company. He was an Oxford history graduate (who considered politics after 1745 as 'thoroughly without interest and even distasteful'), a writer of popular romantic and adventure fiction and a poet. He studied cathedral services and music and was an honorary librarian and reader in palaeography in Durham (to which he moved from Newcastle). He had left his beloved Oxford for the north to become tutor for Noble's sons. He stayed on as Noble's private secretary and in 1897 became the secretary of the newly-formed Armstrong, Whitworth Company.[50] Falkner moved onto the Elswick board in 1902, where he supported Noble on every vote, including relatively small matters on which he held a different opinion.[51] Alfred Cochrane, Noble's son-in-law, became the new secretary. In 1909, another of Noble's sons, John, joined the board. Sir William's grand-nephew and heir, William

Watson-Armstrong had become a director in 1897 and tended to support the Noble faction.

As Noble consolidated power around himself, Rendel's place in the company became uncertain. He had never been a managing director despite his very large shareholdings, but upon Sir William's death he wanted to become vice-chairman. The position went to Cruddas, however, in a vote Rendel saw as a contrivance by Noble. 'It was clear,' he bitterly claimed some years later, 'that Cruddas was put into the position in order to keep me out.'[52] Noble knew the armaments business as well as anyone and he was a major shareholder, but he made virtually all decisions without consulting with or asking the advice of the outside directors, amongst whom Rendel was the most important. Rendel was further estranged from the company by his traditional belief that the firm's directors should own a number of shares commensurate with their responsibilities. He thought that too many Elswick executives held power beyond what was warranted by the number of shares they owned, 'men with little or no stake but with most commanding knowledge and influence'.[53] Rendel's formal power on the Elswick board was certainly incommensurate with his large holdings.[54]

Rendel could change little as an outside director but in 1908 he replaced the ailing Cruddas as vice-chairman and was finally in a position to campaign for reform. In 1910, he was able to replace what he called the 'one-man system' of management with one in which Elswick's lower-level managers were given a louder voice in matters that concerned them. Department heads were renamed 'Local Directors' and met the board regularly to discuss their particular projects and ideas. The treasurer's job was expanded to include the chairmanship of the local directors' meetings. A new treasurer was sought – the Comptroller of the London County Council was approached – to replace John Noble. Sir Andrew Noble continued his sales trips abroad but he relinquished his absolute control over the Ordnance Department. By the end of 1910, Rendel considered that the new management had been 'amicably and unanimously arranged'.[55]

Rendel now turned his attention to enlisting new directors to carry the company into the future. The question of Noble's successor had been in the air since 1900,[56] and had become urgent in 1906 when the 75-year-old Noble fell seriously ill (probably with pneumonia). His sons, Saxon and John, while not without abilities, were not ready to lead a company like Elswick and everyone agreed that Lord Armstrong's heir, Sir William Watson-Armstrong, was quite inadequate for the job. Lord and Lady Armstrong had not had any children and accordingly William Watson, the grandson of Armstrong's sister, was designated heir and in 1889 he added the name Armstrong to his own.[57] Watson-Armstrong became a director in 1897, and no doubt thought that one day he would take over the company. Rendel had had 'warm hopes' for the new recruit, but they were not fulfilled. Watson-Armstrong, said Rendel, 'has a sadly easy command of second rate abilities and this he loves to exercise'. Watson-Armstrong was 'a good, upright, kindhearted excellent fellow', but these were liabilities for the leader of Elswick. 'If he were a worse fellow,' said Rendel, 'he might be less easily made use of by inferior men.'[58] Without Rendel's support, and faced with Noble's ambitions to create his own family dynasty, Watson-Armstrong had no hope. He was forced off the board in 1908 'for the good of the company'.[59] His name is not on the 1911 list of shareholders.

Others too had more ambition than ability. Henry Whithead, who came with the Whitworth amalgamation, put forth his name but he was not considered competent.[60] Sir Henry Alderson thought Stuart Rendel best-fitted to succeed, although he admitted that, at 71, it was probably more than Rendel could undertake. Rendel considered it if only to carry out his reforms, as he would do again in 1910 under similar circumstances,[61] but he realised that the company needed a permanent chairman. Lord Kitchener was considered but, whatever his military reputation, he lacked the knowledge and experience to manage a vast business like Elswick.[62] Admiral John Fisher made the unlikely boast that, in 1905, Noble had told him that all the armaments firms in the country were 'willing to unite under my presidency (and practical dictatorship!)', but this fantasy was simply a part of his machinations at the Admiralty.[63] It may be true that he was offered a directorship at Elswick in 1901 which he turned down when Selborne promised he would soon become First Lord of the Admiralty, but that would have been the extent of his contribution.[64] In the end Noble recovered and the crisis passed.[65]

In 1910, Rendel began a new search for a chairman. He discussed various candidates with his allies on the Elswick Board, Henry Gladstone and Herbert Carington, the latter having been an outside director since 1897. Rendel failed to entice Sir Francis Hopwood with the promise of £10000 a year plus a percentage.[66] Hopwood was a rapidly rising civil servant at the Colonial Office and a member of Edward VII's inner circle. In 1912, he became a Civil Lord of the Admiralty under Churchill.[67] He remained a close friend and advisor to Rendel and helped him in his struggle against Noble and in his search for new directors.[68] Carington did not object to Hopwood, but he thought it essential that they get a man from the War Office or Admiralty since 'relations with them have been on the down grade for some years' and it was therefore important to re-establish strong, friendly links.[69] Rendel was well aware that Elswick's technological leadership remained undisputed, but that her ties with the British state needed attention. He knew that in the twentieth century the successful armaments business would be the one which developed the strongest links to the state. Competition and economies of scale had driven the armaments industry deeper into the political arena. The firm could no longer assume that its technological innovations would suffice to carry the company into the future. Armstrong had always been sensitive to this point. In 1883, when the president of the Ordnance Select Committee was considered as a director, he told Noble that 'I would rather see him president of the O.S.C. than a director of our Co. After being a shareholder in Elswick he would be [our own?] man at the head of the Committee.'[70]

Elswick also needed to strengthen its internal management. Elswick, said Rendel, had 'an efficient technical staff and plenty of work' but was 'gravely defective in the lead of its affairs'. He explained to Hopwood that, 'had the business continued to rest on the old notions of provision of the best goods on the best terms, our steersmen would have proven as efficient as our staff. But the business in War Material has left the old lines. It is more ministered by political and diplomatic influences and complicated by Rings and Combines. It demands for its guidance quite new & special qualities ... the aid of men of wide scope of experience of men & affairs'.[71] Rendel himself remained Elswick's most effective bridge to the British state. As Falkner said, Rendel could 'certainly intervene [at the Foreign Office] with much more weight than anyone else'.[72]

Rendel succeeded in bringing onto the board three men with administrative and financial experience and close ties to the state apparatus. The first was Sir George Murray, a man with a long and illustrious career. He began in the Foreign Office in 1872 and moved in 1880 to the Treasury, where he was private secretary to Gladstone between 1892 and 1894. He was chairman of the Board of Inland Revenue in 1897 and became the permanent administrative and financial secretary to the Treasury in 1907. He was chairman of a committee which recommended further reductions in orders from the Royal Ordnance Factory and an increase in the production of armaments by private enterprise.[73] He had experience in finance and business as a director of a bank and a railway company, an ability to master complex financial and administrative problems, and great physical and constitutional endurance.[74] Like Rendel, Murray was an old Gladstonian.[75] He became a director in 1912.

A year later, Elswick obtained its Admiralty man when Sir Charles Ottley joined the board. Ottley had entered the navy in 1871 and rose rapidly thanks to his technological knowledge of submarine mines. As a director of the Imperial Ottoman Docks, Arsenals, and Naval Construction Company, Ottley was an important link between Elswick and the Turkish navy. Furthermore, Ottley had been naval attaché at British embassies from St Petersburg to Tokyo and thus had valuable knowledge about, and access to, many foreign naval departments. But his greatest appeal was his close association with the British military and naval establishments. In 1905, he had joined the staff of the new Committee of Imperial Defence (CID) and in 1907 had succeeded Sir George Clarke as its secretary.[76] Although his knowledge of foreign markets was useful, in Rendel's view, Ottley was needed 'more especially at home'.[77]

The third man whom Rendel brought to Elswick just before the war was Sir Percy Girouard. Girouard was a railway engineer and colonial administrator. He had been traffic manager at the Woolwich Arsenal and director of the Sudan railways which had contributed so much to Kitchener's conquests. He made a similar contribution to the South African campaign. He succeeded Sir Frederick Lugard as Governor of Northern Nigeria and supervised the construction of railroads throughout that colony.[78] His career reached an impasse in 1912 and, as Hopwood explained to Rendel, he was 'very keen to become connected with Armstrongs'.[79] He joined the firm in 1913.

These men revitalised Elswick's financial and administrative management. More important, they strengthened the close, cooperative ties between the company and the British state which, in the era leading up to the First World War, had become as important as developing new technologies. Perhaps Elswick's relations with states (who were, after all, its customers) had always been as important as developing new weapons. The interplay between technological innovation and the social and political environment of the day cannot be easily separated into causes and effects and the debate will continue between those who see technological innovation as an expression of an existing social–political system and those who insist that technological evolution has an inner momentum which carries society and people's minds along with it.[80]

The dramatic leaps in heavy gun technology which began with Sir William Armstrong's little 3-pounder gun of 1855 seem to have been associated with some political struggle, strategic crisis or economic problem, and men like Armstrong were no doubt driven by the profit motive. On the other hand, the technological development of armaments was an evolutionary process which even a powerful political figure like Lord Palmerston had to wait upon. Moreover, it is not entirely certain that Armstrong, at least in 1854–5, was motivated only by thoughts of profit; he also had a desire to solve an engineering problem. This question is explored in the concluding chapter, but here it should be noted that, as Elswick grew into a large company and military and naval technology surpassed 'civilian' engineering after 1880,[81] the quest for profits and markets became absolutely essential to the survival of the company. At the same time the survival of Elswick came to be seen by the British government as essential to the survival of Britain. This was the basis of partnership between Elswick and the British Foreign Office and the broader political–industrial complex which developed in Britain between 1880 and 1914. It was a goal towards which Sir William Armstrong had worked since the day in 1854 when he was introduced to the secretary of state for war. The relationship between Elswick and the British state had taken 60 years to establish itself, but by 1914 it had become a fundamental component of the new age of warfare that was about to commence.

Clive Trebilcock referred to a 'special relationship' between the British state and the armaments companies.[82] He also referres to the relationship as a partnership in which the state was the senior partner.[83] His definition of the state, however, is little more than the procurement departments of the War Office and the Admiralty. This history of Elswick has employed a broader conception of the state which included the Cabinet, Parliament as well as the army and navy bureaucracies and concluded that Elswick was the partner which continually pressed the others with its technological innovations and its international business. Before 1880, the Cabinet arm of the British state maintained a system in which armaments production would spring into action only at the moment of war. The level of armaments spending fluctuated according to purely political–diplomatic decisions. Thus, during peacetime – the 1860s and 1870s for example – the government reduced armaments expenditures as far as possible, and were reluctant to promote technological evolution. Thus, in 1862, the government reduced armaments spending, cancelled the Elswick contract, returned to cheaper muzzle-loaders and allowed the Royal Navy to languish.

The navy survived without Armstrong between 1862 and 1880 (supplied by the War Office with modified, cheaper versions of Armstrong guns), but so too did Elswick survive without the British navy and, indeed, the company flourished in the global markets of the 1860s and 1870s. During the 1880s, technological and political circumstances altered. Production of weapons could no longer begin when war was declared. The concept of deterrent entered a period of renewal.[84] Economic depression and the growing restlessness of the working classes in the era of expanded democracy (and an Independent Labour Party) meant the government had to consider social and economic factors in their policies and strategic decisions which affected the armaments industry.[85] In the 1880s, Elswick had 15 000 workers; in the 1890s, there were 25 000 workers and more than £2 million capital invested. In the company's eyes the distinction between the interests of Elswick and those of the nation was blurred. Elswick, said Rendel, was more than just a private factory; it was 'our second arsenal' and 'of public importance'. Its sufferings were those of the nation.[86] This idea was conveyed to the British public and the government by the lengthy description and illustrations of its technological immensity that was published by the *Pall Mall Gazette* in a special 'extra' edition in 1884 – that Elswick was a *national* arsenal.[87] Rendel conceded that political considerations would, from time to time, bend this principle – like the order for a cruiser going without tender to an Irish yard (regretfully Orange, he said) – but this, he said, should be exceptional.[88] Armstrong too wanted Woolwich and Elswick to act 'in concert'.[89] By this he meant that the government should bear the costs of research and development and leave production to private firms.[90] He had been consistent in the belief since his first meeting with the Duke of Newcastle in 1854 and he had reaffirmed it in his letter of resignation as Engineer of Rifled Ordnance in 1863, in which he invited the government to remember that he would 'at all times be ready gratuitously to render the War Department any assistance'.[91]

After 1884, strong bonds developed between the steel and ship-building industries and the government which translated into an expanded naval armaments industry.[92] Naval scares seemed to come at the end of the business cycle in the steel and ship-building industry.[93] By 1902, *The Economist* considered that 'interesting and important as are the new Admiralty contracts from a naval and military point of view, they are scarcely of less importance from an economic standpoint' and, given the economic stagnation in that vital industry, concluded that 'the allocation of naval contracts among private builders at this time is welcomed as a great economic relief'.[94] The system of armaments loans had taken the business to new heights and fears of general bankruptcy crept into the minds of a few. *The Economist* said, 'the world is overloaned and overarmed', and described the greatest amalgamation of the era before the First World War ('a great and growing evil') in which armament manufacturers, banking interests and diplomacy combined to sell their wares to the world.[95] By 1914, it had become difficult to distinguish between the business of Elswick armaments and the business of the British state. Whether or not it was recognised by the British government in 1914, Elswick had become an engineering enterprise of national importance, fulfilling the vision held by Sir William Armstrong in 1859.

Notes

1. P.L. Payne, 'The Emergence of the Large-scale Company in Great Britain, 1870–1914', *Economic History Review*, 20 (1967), pp. 53–4 places the

Armstrong company eleventh out of 52 in the post-amalgamation period. Biographical information on the 31 members is from a wide variety of sources, but the main ones are the *Dictionary of National Biography* (*D.N.B.*), *Who Was Who*, *Who's Who*, Burke's *Peerage*, *Debrett's* and the *Dictionary of Business Biography* (*D.B.B.*).

2. J.D. Scott, *Vickers: A History* (London: Weidenfeld & Nicolson, 1962) and David Dougan, *The Great Gun Maker: The Story of Lord Armstrong* (Newcastle: Graham, 1971) state that Armstrong withdrew from the management of the firm after 1880, but the evidence indicates that he remained active in all major decisions right up to his death.
3. James Hazlett, Edwin Olmstead and M. Parks, *Field Artillery Weapons of the Civil War* (Newark: University of Delaware Press, 1983), p. 197; *The Times*, 27 Feb. 1865, p. 12 and 28 Feb., p. 12.
4. Frederick Manning, *The Life of Sir William White* (London: John Murray, 1923), p. 103.
5. Bernard Broomhead, 'Royal Commission on Warlike Stores', q. 9941.
6. Sir W.G. Armstrong, Mitchell & Co., *Minute Book*, Armstrong Archives, T&W, 130/118; Vavasseur, evidence to the 'Royal Commission on Warlike Stores', qs 9558–64; Sir Sydney Eardley-Wilmot, *An Admiral's Memories* (London: Sampson Low, 1927), pp. 122–3.
7. William H. McNeill, *The Pursuit of Power: Technology, Armed Force, and Society Since A.D. 1000* (Chicago: University of Chicago Press, 1982), p. 279.
8. Rendel, *Personal Papers*, pp. 221, 282.
9. *The Times*, 23 Sept. 1884, p. 5.
10. Rendel to Grant-Duff, 7 Oct. 1884, N.L.W., Letters and Papers of Lord Rendel, no. 382. Stuart had acquired the shares of his brother George when the latter went to the Admiralty.
11. Armstrong to Rendel, 24 Dec. 1896, T&W, 31/3684. The issue arose again at this time no doubt as a result of the discussions leading to the amalgamation with Whitworth.
12. Ibid.
13. White evidence, 'Second Report of the Select Committee on Naval Estimates', *Parl. Papers* XIII (1888), qs 7566–81.
14. White to Select Committee on Naval Estimates, Second Report, *Parl. Papers* XIII (1888), q. 8016.
15. Ibid., q. 7582.
16. Ibid., qs 7585–6; Manning, *Life of White*, chs 8–12.
17. White, evidence to 'Royal Commission on Warlike Stores', q. 7589.
18. This condition was dropped when it was used by political and bureaucratic opponents of White's return to the Admiralty. White demanded a substantial increase in the salary of the navy's chief constructor, to which the Treasury had to agree, given the political pressures created by the naval scare of 1884–5. However, it was still much lower than his Elswick salary (Manning, *Life of White*, p. 178).
19. There is no evidence as to why White did not obtain a seat on the board. It may be a reason for his leaving Elswick.
20. *Economist* (4 Jan. 1913), p. 16. Smith was not a director.
21. *Who Was Who* (1897–1916).
22. *D.N.B.* (1901–1911); *Who's Who* (1902); O.F.G. Hogg, *The Royal*

Arsenal: Its Background, Origin and Subsequent History, Vol. II (London: Oxford University Press, 1963), pp. 1039, 1112.

23. *Who Was Who* (1936); Sir John Adye, *Recollections of a Military Life* (London: privately published, 1895), p. 289.
24. *Hansard*, 6 Jan. 1881, cols 75–84.
25. Grahame Nelmes, 'Stuart Rendel and Welsh Liberal Political Organization in the late-Nineteenth Century', *Welsh History Review* (1978–9), pp. 468–85.
26. Henry Gladstone to Rendel, 6 Oct. 1902, T&W, 31/3644.
27. Stuart Rendel, *The Personal Papers of Lord Rendel*, ed. F.E. Hamer (London: Ernest Benn, 1931). See also his correspondence with Hart on this in T&W, 31/3980–/3990.
28. John Wilson, *CB: A Life* (London: Constable, 1973), p. 594.
29. *Hansard*, 9 Feb. 1888, cols 14–23. For the details of the 1886 campaign, see Dougan, *Great Gun Maker*, pp. 156–63.
30. *Hansard*, 27 May 1862, cols 49–50.
31. He did not enjoy being chairman of various railway committees and turned down Gladstone's offers of chief whip or postmaster-general. His dinner parties were famous and his most enjoyable position was chairmanship of the National School of Cooking, which he held from 1874 until 1903. (*D.N.B.*, 1907).
32. H. Gladstone to Rendel, 17 July 1907, T&W, 31/3692.
33. Burke's *Peerage* (1970); *Who's Who* (1902).
34. *Hansard*, 17 March 1879, cols 999–1007; 3 April, cols 245–51.
35. W.J. Reader, *Imperial Chemical Industries: A History*, vol. 1 (London: Oxford University Press, 1970), pp. 131–2.
36. Michael Stenton, *Who's Who of British Members of Parliament*, vol. 1 (Hassock, Sussex: Harvester Press, 1976).
37. *Debrett's* (1881).
38. *D.N.B.* In 1887, some London ratepayers alleged that Hogg and others had used their positions for personal gain.
39. Hamilton Rendel to Stuart Rendel, 28 Jan. 1885, T&W, 31/2622.
40. *Hansard*, 27 May 1889, col. 1083. Cruisers, he said, represented the traditional offensive strategy of the British navy and ought to be maintained.
41. Noble to Rendel, 24 March 1885, T&W, 31/5280.
42. R.W. Seton-Watson, *A History of the Rumanians* (London: Archon, 1963), p. 344.
43. T&W, 31/5278, /5280, /2677, /2678, /2679.
44. Seton-Watson, *History of Rumanians*, pp. 344, 361–3.
45. Hamilton Geary to Noble, 21 Dec. 1881, T&W, 31/5278.
46. Ibid.
47. Dyer to Rendel, 28 Nov. 1883, T&W, 31/3242.
48. The issue does not appear in *Hansard*.
49. *Who Was Who* (1941–1950). Nickel was an important alloy in armour plate.
50. *D.N.B.* (1931–1941).
51. Falkner to Rendel, 15 Jan. 1904, T&W, 31/3645.
52. Rendel to Falkner, 1 Feb. 1905, T&W, 31/7055.
53. Ibid.
54. In 1906, the managing directors held the following number of shares:

Andrew Noble, 270 125; Saxon Noble, 19 500; John Noble, 19 500; William Cruddas, 21 550; Henry Swan, 63 000; Josiah Vavasseur, 50 000; Percy Westmacott, 21 000; Augustus Albini, 2000; William Watson-Armstrong, 62 000 (plus 152 000 'with others'); while the outside directors consisted of Stuart Rendel with 230 000 shares, Henry Carington with 2000 and Henry Gladstone with 10 000 (plus 4000 'with others'). (Data from Sir W.G. Armstrong, Whitworth & Company, 'List of Holders of 1000 Ordinary Shares and Above', 5 Jan. 1906, T&W, 31/6602.)

55. Rendel to Sir Francis Hopwood, 24 Dec. 1910, T&W, 31/7489. See also Noble to Rendel, 28 July 1911, T&W, 31/6712 and Noble to Rendel, 2 Dec. 1911 in Rendel, *Letters and Papers*, N.L.W., no. 1517. These conflicts are described in Scott, *Vickers*, pp. 88–94 and in C. Treblicock, *The Vickers Brothers: Armaments and Enterprise 1854–1914* (London: Europa, 1977), p. 146. A useful analysis is provided by R.J. Irving, 'New Industries For Old? Some Investment Decisions of Sir W.G. Armstrong, Whitworth & Co. Ltd., 1900–1914', *Business History* 17 (July 1975), pp. 160–63.
56. Falkner to Rendel, 4 Jan. 1901, T&W, 31/3597.
57. Armstrong's title died with him in 1900, but in 1903 Watson-Armstrong was created the first Lord Armstrong of Bamburgh and Cragside (the names of the two estates owned by Armstrong).
58. Rendel to Falkner, 1 Feb. 1905, T&W, 31/7055.
59. T&W, 31/7554. See also 31/7546 and 31/7551.
60. Henry Gladstone to Rendel, 12 Feb. 1906, T&W, 31/3668.
61. Rendel to Sir Francis Hopwood, 24 Dec. 1910, T&W, 31/7489.
62. Sir Arthur Godley to Rendel, 8 and 9 Feb. 1906, N.L.W., Rendel, *Papers*, nos 252, 253.
63. Philip Magnus, *King Edward VII* (Harmondsworth: Penguin Books, 1964), p. 451.
64. Peter Kemp, 'From Tryon to Fisher: The Regeneration of a Navy', in Gerald Jordan (ed.), *Naval Warfare in the Twentieth Century: Essays in Honour of Arthur Marder* (London: Croom Helm, 1977), p. 26.
65. Falkner became the next chairman when Noble died in 1915. Rendel died in 1913.
66. Rendel to Hopwood, 17 July 1910, T&W, 31/7484.
67. *D.N.B.* (1941–50).
68. See, for example, Hopwood to Rendel, May 1910, T&W 31/7480; Rendel to Hopwood, 17 July 1910, 31/7484 and 2 Dec. 1910, 31/7489.
69. Carington to Rendel, 15 Feb. 1909, T&W, 31/6191.
70. Armstrong to Noble, 3 March 1883, T&W, 31/3133.
71. Rendel to Hopwood, July 1910, T&W, 31/7485.
72. Falkner to Rendel, 4 Nov. 1911, T&W, 31/7511.
73. 'Report of the Government Factories and Workshops Committee', *Parl. Papers* X (1907).
74. *D.N.B.* (1936).
75. Roy Jenkins, *Asquith* (London: Collins, 1965), p. 160.
76. Clarke joined Elswick in 1916.
77. Rendel to Hopwood, 30 June 1912, T&W, 31/7447. Ottley's biographical information is from *D.N.B.* (1931–40) and Philip Noel-Baker, *The Private Manufacture of Armaments*, vol. 1 (London: Victor Gallancz, 1937), p. 72, fns 1 and 3. For details of Ottley's work on C.I.D., see F.A. Johnson,

Defence by Committee: The British Committee of Imperial Defence 1885–1959 (London: Oxford University Press, 1960), pp. 79, 86, 98–9.

78. *D.N.B.* (1931–1940).
79. Hopwood to Rendel, 23 April 1912, T&W, 31/7507.
80. The tension between these two perspectives is apparent in McNeill, *Pursuit of Power*. At one moment (p. 278) he suggests that 'it is impossible to say' how much weight should be assigned to 'technological innovation as an autonomous element' in the interplay among technology, domestic politics and international rivalries, but a little later (p. 294) he seems confident that 'the rush of new technology that cascaded upon the Royal Navy after 1884 ... began to get out of control itself'. See also his remarks on pp. 298 and 306.
81. An argument developed by Clive Trebilcock, 'Spin-off in British Economic History: Armaments and Industry, 1760–1914', *Economic History Review* 22 (1969), pp. 474–80. McNeill argues a similar point with his concept of 'command technology' which he sees emerging in Britain after 1880. See *Pursuit of Power*, pp. 278–85.
82. Trebilcock, 'A "Special Relationship" – Government and the Cordite Firms', *Economic History Review*, 19 (1966), pp. 364–79.
83. Trebilcock, 'Legends of the British Armaments Industry 1890–1914: A Revision', *Journal of Contemporary History* 5 (1970), pp. 3–19.
84. 'Peace through strength' was not challenged until 1905, when Campbell-Bannerman argued that armaments themselves increased the risk of war. See his speech in Alan Bullock and Maurice Shock (eds), *The Liberal Tradition: From Fox to Keynes* (Oxford: Clarendon Press, 1956), pp. 237–8. Sir Edward Grey, Britain's Foreign Secretary from 1905 until 1916, was ambivalent both before and after the war, but ultimately seems to have believed that human psychology, not weapons, cause war. See his speech in ibid., pp. 239–41; his speech to the Manchester Chamber of Commerce, *The Times*, 4 Feb. 1914, p. 10; his speech in Parliament, *Hansard*, 18 March 1914, cols 2183–4 and his postwar memoirs: Viscount Grey, *Twenty-Five Years 1892–1916*, vol. 1 (New York: Frederick Stokes, 1925), pp. 89–90.
85. Arthur J. Marder, *The Anatomy of British Sea Power: A History of British Naval Policy in the pre-Dreadnought Era, 1880–1905* (New York: Alfred Knopf, 1940), pp. 33, 36–7.
86. Stuart Rendel, 'Memorandum on Elswick & Woolwich Guns', 1886, T&W, 31/5288.
87. 'Britain's Supplemental Arsenal: At Elswick', *Pall Mall Gazette*, 1884. A copy is in T&W, 31/5288.
88. See Rendel to Lord Spencer, 20 Oct. 1894, and Spencer's reply the next day: Rendel Letters and Papers (N.L.W.), nos 733 and 734. Some armaments orders had always been given for purely political reasons, but Armstrong and Rendel to a large extent had created their own political significance.
89. Armstrong to Noble, 3 March 1883, T&W, 31/3133.
90. Armstrong, evidence to the Royal Commission on Warlike Stores, *Parl. Papers* XV (1887), q. 9510.
91. 'Sir William Armstrong: Correspondence', *Parl. Papers* XXXII (1863), p. 221. Armstrong had a strong sense of patriotism, and was ready to help the government with his expertise by serving on the Coal Commission.

92. The new links between industry and government were forged by the 'Report of the Committee on the Organization and Administration of the Manufacturing Departments of the Army' (Morley Committee), *Parl. Papers* XIV (1887). See also the 'Letter from Certain Steel Manufacturers of Sheffield to the Chairman of the Select Committee on Army Estimates Together with His Reply', *Parl. Papers* LXXXI (1888), pp. 529–31. The manufacturers were Firth, Vickers and Cammell; the chairman of the committee was Randolph Churchill. See also *Hansard*, 16 March 1885, cols 1298–9.
93. See all the works of Arthur Marder, especially *Anatomy of British Sea Power*, ch. 3, 'The Armaments Industry and Navalism', esp. pp. 34–7; Sidney Pollard, 'Laissez-Faire and Shipbuilding', *Economic History Review* 5 (1952), pp. 98–115; Oron J. Hale, *Publicity and Diplomacy: With Special Reference to England and Germany, 1890–1914* (New York: D. Appleton-Century, 1940); A.J.A. Morris, *The Scaremongers: The Advocacy of War and Rearmament, 1896–1914* (London: Routledge & Kegan Paul, 1984).
94. 'Industrial Aspects of the Admiralty Contracts', *The Economist*, 12 April 1902, p. 564.
95. 'Overloaned and Overarmed', *The Economist* (24 May 1913), pp. 1274–5.

Conclusion

I

The growth of Elswick into a vast industrial enterprise is one exception to the consensus amongst historians that 1860–1914 was a period of relative industrial decline in Britain brought about by the poor entrepreneurial performance of its industrial managers, and especially those whose business relied on the technology of iron and steel. There is no doubt that after 1860 Britain gradually lost its overwhelming industrial pre-eminence of the early nineteenth century and became one of three industrial giants in 1914. The debate is about why this relative decline occurred, with the traditional answer pointing to the failings of Britain's entrepreneurs and senior managers. Two mid-twentieth-century historians summed up one side of the debate by concluding that, 'if a business deteriorates, it is of no use blaming anyone except those at the top'.[1] This view leads to a great emphasis on personalities and decisions inside the company over external historical and contextual factors, and is exemplified by a more recent study of Elswick which concludes that 'the weakness lay at the top'. The author is quite radical in stating that Elswick's 'Contracts with the War Office, Admiralty and Government and, for the pioneer of large-scale overseas sales of armaments, even more complicated relations with a large number of foreign governments', were excluded because 'such matters would be a distraction in a business history'.[2]

The other approach, and the one taken by the present work, is to observe the interplay between internal, external and historical forces in the origin, development and decline of Elswick. Like political history, business history as the biography of great men has its limitations. The motives, abilities and decisions of a 'vital few' are less important than they appear, in that they themselves are the product of numerous interacting external forces, opportunities and traps, none of which provides a sufficient explanation of events. Elswick grew out of the interactions between technological innovation and the political circumstances.

Great industrial enterprises succeed or fail for many reasons, some internal to the firm, others external. The two important ingredients in the success of Elswick were, first, the technological innovations of Sir William Armstrong and, secondly, the sensitivity of the company to the political context, so vital to the armaments business at any given time and place. Thus Elswick's history must highlight two fronts: the internal history of technological innovations and management, on the one hand,

and the external forces in the world beyond. Thus the history of Elswick can be told in several ways and from different perspectives, and through the eyes of several protagonists: upper management, workers, the surrounding community and its customers. At the broadest level Elswick's history is part of the grand narrative of the imperialism of industrialism capitalism, the relentless spread of commerce and technology which began with the Crimean War. Thus Elswick is embedded in modern global history. Armstrong's technological innovations brought together traditional metallurgical techniques with the new areas of knowledge of chemistry and electricity. His first 3-pounder gun was developed in 1855 because it was technologically and politically possible to produce such a weapon at that moment. Armstrong's personal historical importance is based on his success in applying his engineering skills to the spirit of the moment and adapting to the changing context as that spirit evolved.

The social and political context involved external factors far beyond Armstrong's command. It was his great strength that he did not attempt to change the world to suit his business, but adapted his businesses to suit the changing world. On the other hand, while Armstrong's friendship with James Rendel opened the door to the British government, he still had actually to construct the weapon himself. In other words, the Armstrong built-up artillery was a product of both the individual talent of Sir William Armstrong and the technological, political and international context in which he lived. Without that passing external context, Armstrong would have no doubt continued to develop civilian engineering. This interplay between public context and individual effort remained important in the development of British armaments and the evolution of Elswick. Again and again we have seen how armaments technology, the armaments business and domestic and international politics moved together.

This history of Sir William Armstrong and the development of Elswick into an industrial enterprise of global dimensions is thus in one way a narrative of the rise of a Victorian engineering entrepreneur and businessman: the story of a man who combined a natural ability to synthesise past engineering innovations into a breakthrough technology, and an astute business sense to transform that technology into a new industry. While developing his new guns, Armstrong was driven by two forces. The first was a relentless determination to pursue technological innovation in any new field for the sake of technological innovation itself. Armstrong remained primarily interested in technology and science. At a deeper level, this emotional commitment to engineering was fuelled by the unbounded Victorian optimism in the benefits of technological progress. This general embracing of science and technology in the nineteenth century made cultural heroes of men like Armstrong, and made possible the enormous growth of Armstrong's company.

Armstrong left little evidence elucidating his personality, inner thoughts and motives or his family life. Virtually all his private letters to Rendel and his public letters to *The Times* are concerned with armaments engineering and the armaments business. His patriotism is obvious and may have affected his business decisions, at least in the very early years, but one can only speculate about how much relative importance Armstrong attached to the personal satisfaction of making technological innovations compared to, say, recognition and adulation from his peers and various nations (of which his peerage was no doubt the most important) and the defeat of competitors, especially Whitworth, or the accumulation of the enormous profits which his innovations brought. As one pieces together Armstrong's technological and business adventures, a sensation emerges which suggests that, after his great technological innovations made between 1855 and 1865, his attention turned more towards managing the business of Elswick. After Elswick became a limited liability company in 1882, and as it expanded into a vast, complex organisation in the 1890s, the fluctuations in demand associated with the business cycle required that greater attention be given to managerial and organisational matters than to technological innovation. The high capital costs of the armaments business necessitated that Elswick transform itself from a partnership into a joint-stock, limited liability company. Armstrong moved with apparent ease from being a traditional mid-Victorian entrepreneur to the condition of a modern late-Victorian corporate capitalist.

This work has not cast Elswick as a case study in the entrepreneurial vision and managerial performance of British late Victorian entrepreneurs largely because such an approach has shaped virtually every other study of the company, and it excludes far too much of the company's rich history and other questions such as the nature of technological innovation and the larger contexts in which the company existed. The intention here has been to discover both individual motives and aims of the company's managers, intersected by external circumstances, to uncover the workings of the global armaments business and the military–industrial–political complex which emerged in the nineteenth century.

The entrepreneurial thesis has two fundamental weaknesses, one general and one particular to Elswick. First, the general assumption that all businesses rise or fall exclusively on the quality of their senior executives, while often stated, is seldom demonstrated. It ignores far too many fluctuating external contextual factors which determine a company's development, from market demand, transport and communication to government policies and international relations. Just as technological change is far more complicated – and interesting – than the internalist theory allows, so entrepreneurial and managerial decisions, while important, are not the only variable that must be taken into account in analysing the development of a business, an industry or an economy. An overemphasis on an internalist model of management is a distortion in that it entails an extremely narrow definition of technological innovation and entrepreneurship, and cuts off important external realities. The entrepreneurial model makes for history which is too narrow and simplistic.

The second, particular, weakness of the entrepreneurial failure thesis is that it fails when applied to Armstrong Whitworth. There is simply not the evidence to argue that Elswick in any of its particular forms was a company in entrepreneurial decline. The archival records on this issue are abundant but contradictory. The letters of Stuart Rendel provide an important subjective insider's analysis of the company's managerial troubles, and are the source upon which the entrepreneurial failure argument rests, while the objective financial record of the company – as 'objective', that is, as any human construction, whether a balance sheet or a new artillery piece or warship, can be – provides the best evidence that the company remained efficient, rational and profitable. Rendel criticised Elswick's executive directors for paying themselves high salaries by unethically if not illegally burying the payments in company accounts, rather than using dividends to distribute profits. Such a policy would significantly benefit only Rendel and Andrew Noble, the two major shareholders.

It was his battle for higher dividends that led the aggressive and determined Rendel into exaggerated claims about the weak financial management and relative decline of Elswick as a whole. Rendel looked back to the traditional practice in which company executives were also large shareholders, but the new directors could not afford to buy large numbers of shares, and even a company the size of Armstrong Whitworth was insufficiently capitalised to allow Rendel to liquidate his shares. The capital markets of the time were simply unable to allow for the transfer of ownership. In the new era, corporate managers could not also be major shareholders, and some other way of paying them had not yet developed. This was the financial dilemma which the company struggled to overcome; it was not a sign of declining managerial competence.

It is well to note that, throughout his campaign for managerial reform and renewal, Rendel saw no decline whatsoever in Elswick's technological leadership and productive efficiency. Nor does the company financial record suggest anything less than a results-oriented management which made the rational strategic decisions. Rendel's concerns were a reaction to broad changes in the business context that was unfolding in the early twentieth century in which the traditional intimate relationship between shareholder capital and management of large firms was separating into two related but distinct elements in the modern structure of businesses. Rendel's anxieties about the long-term financial strength of Armstrong Whitworth indicate the increasingly subjective nature of revenues and profits owing to the several ways they could be calculated in large diversified companies with outside investments and high research and development expenditures. Rendel was quite right in highlighting the importance of sound financial management, and the debate he prompted focused attention on adjusting the balance between dividends and salaries on the one hand, and capital reserves, research and development expenditures and share price on the other.

That Armstrong Whitworth was a growth company which paid high dividends is clear evidence of the extraordinary financial strength of the company. The return on capital averaged a fully-earned 11 per cent between 1889 (the year of the Naval Defence Act) and 1897 (the year of the amalgamation with Whitworth), and 15 per cent per year over the following decade. Profits fell to 10 per cent in the general cyclical decline of 1908–10, but recovered to 15 per cent in 1910–14. The company's highest profits came from armaments, where profits averaged 20 per cent between 1902 and 1912, with the highest profits coming from its Openshaw armour plate works, the acquisition of which was the reason for acquiring Whitworth in 1897, and from the gun steel made in the Elswick Steel Works, the major strategic investment which immediately followed the breakthrough in steel manufacturing made in the 1870s, and which was the technological foundation of the rapid growth of the company in the 1880s. Guns and armour remained the oldest but still most profitable section of the armaments industry and Elswick was fully invested in those two essential areas of the business.

The volatility of the warship and ordnance markets made the Elswick Shipyard and the Ordnance Department less profitable, but essential components of an armaments business to be maintained for the time when the inevitable shift back to ships and guns would occur. The financial structure and industrial base were sound and the company was positioned to weather market fluctuations in armaments, or if necessary expand its commercial ship building or car production should one or both of those businesses become more profitable. Despite profit levels of less than 10 per cent in civilian products, Elswick retained them as part of a sound diversified industrial strategy.

The prosecution's case against Elswick's entrepreneurial leadership rests on the claim that the company allowed Vickers to overtake it within 10 years of the latter company entering the armaments business in 1887. Yet the growth of Vickers was the result of British government policy to enlarge the private armaments industry and increase price competition within that industry and shift the financial risks of a volatile armaments market to private industry. Vickers grew rapidly in the 1890s, as did Elswick, and for reasons associated with British imperial and foreign policy. The government rejected rationalising armaments production because of its ideological commitment to competition. Rationalisation would have to await the post-First World War era. The Naval Defence Act revealed the degree to which the government depended on Elswick for guns and mountings, which violated its political instinct for competition.

Elswick's management is accused of moving too slowly and being outmanoeuvred by Vickers in the purchase of Maxim-Nordenfeldt and the Barrow naval yard in 1897, of passing up an opportunity to join Vickers in investing in the cash-starved armour plate company of William Beardmore, half of which Vickers bought in 1902, of failing to seize the opportunity to invest in the American Dockyard and Ordnance Company, and of making an unprofitable expansion of manufacturing plant in Italy. Yet Elswick made the rational decision to build a new naval yard at Elswick rather than buy at Barrow-in-Furness (which gave Vickers its own problems); owned a licence to manufacture Gatling machine guns and had no need or opportunity to buy Maxim-Nordenfeldt; escaped being fleeced on the one hand by the overpowering person of William Beardmore, whose directorship at Vickers was used only as a means to channel Vicker's profits into his own company, and on the other by the vague and dubious schemes at the American Dockyard Company; and owned plant in Italy which was profitable and represents an early case of an industrial company investing in its own plant in another country, thus making Elswick one of the first multinational companies. All in all, Elswick management before the war was rationally cautious and stayed focused on its business plan of producing high-quality products (whether armaments or cars) and maintaining close relations with the government. There is no evidence to support the anti-entrepreneurial case.[3]

II

The predominant theme of this history has been the relationship between Elswick and various governments throughout the world, but especially the British state. Technological innovations by private entrepreneurs like

Armstrong naturally affected government demand for weapons, and influenced the structure of international relations. The importance of establishing and maintaining access to the highest levels of government, and cultivating supporters within the naval bureaucracies, was seen most clearly in Britain, China and Italy. The British government was the most important factor in the growth of Elswick between 1859 and 1863 and again after 1889. During the first period, the British government was driven to Armstrong by strategic considerations (the perceived threat from France), but in the 1880s economic factors were included in government calculations. The state realised that national security depended upon the new technologies which were developed by private companies and it quickly moved to establish a closer relationship with firms like Elswick. The granting of a peerage to Sir William Armstrong in 1887 is a symbol of the great importance Elswick had come to assume in the defence of Britain and the maintenance of her place in the hierarchy of the world's nations.

The Northbrooke naval programme of 1884–5 was the first to take into account the economic situation of the private ship-building trade and to acknowledge that Britain's military and naval mobilisation was in large part an industrial mobilisation. The British state came to rely upon the technological knowledge and industrial capacity of firms like Elswick, while Armstrong recognised that the company's share in British armaments orders did not result only from its technological innovations and industrial strength but also from cultivating strong social bonds between the company and the state. After 1900, as the original owners and managers ended their careers, the company lost some of its close ties with the government, and it became Stuart Rendel's goal to re-establish them. He succeeded in this, just before his death in 1913, and in 1914 Elswick was poised to move into the future, confident that it would be much like the past.

Until 1914, Elswick held a premier place in the British naval–industrial complex. It supplied the government with three directors of naval construction, and offered much advice to various War Office committees. The company had originated in that context, it never refused the government its cooperation and expertise, and it always sought to accommodate the War Office, the Admiralty, Cabinet and Parliament in every way. The government was less accommodating and perhaps had a too narrow or one-sided view of the kind of relationship it ought to have with the private manufacturers. Elswick remained committed to supplying Britain with the best defence possible and acceding to the government policies and demands for technological leadership and patriotism. These had been among the founding impulses of Sir William Armstrong in 1854, and they remained a guiding principle. In 1914, Elswick was a financially sound, technologically innovative and well-managed armaments company.

Notes

1. T.H. Burnham and G.O. Hoskins, *Iron and Steel in Britain 1870–1930: A Comparative Study of the Causes which Limited the Economic Development of the British Iron and Steel Industry between the Years 1870 and 1930* (London: Allen and Unwin, 1943), p. 271.
2. Kenneth Warren, *Armstrongs of Elswick: Growth in Engineering and Armaments to the Merger with Vickers* (Basingstoke: Macmillan, 1989), pp. 253, 255.
3. Warren, who takes a radical internalist, management-is-everything approach, is ambivalent on this point. See *Armstrongs*, chs 7, 10, 11, 14–16. For a critique of the case against the entrepreneur, see L.G. Sandberg, 'The Entrepreneur and Technological Change', in Roderick Floud and Donald McCloskey (eds), *The Economic History of Britain Since 1700*, vol. 2: *1860 to the 1970s* (Cambridge: Cambridge University Press, 1981), pp. 99–120; Donald McCloskey and Lars Sandberg, 'From Damnation to Redemption: Judgements on the Late Victorian Entrepreneur', in Donald N. McCloskey (ed.), *Enterprise and trade in Victorian Britain* (London: George Allen & Unwin, 1981), pp. 55–72, Donald N. McCloskey (ed.), *Essays on a Mature Economy: Britain After 1840* (Cambridge, Mass.: Princeton University Press, 1971) and Donald M. McCloskey, *Economic Maturity and Entrepreneurial Decline: British Iron and Steel, 1870–1913* (Cambridge, Mass.: Princeton University Press, 1973). See also P.L. Payne, *British Entrepreneurship in the Nineteenth Century* (London: Macmillan, 1974) and François Crouzet, *The Victorian Economy*, trans. A.S. Forster (New York: Columbia University Press, 1982).

Epilogue

I

In his postwar memoirs, David Lloyd George recalled that, while the Royal Navy was fully prepared for war against the mighty continental power of Germany, the small professional British army had been structured for imperial police duties, with the result that 'the outbreak of war found this country totally unprepared for land hostilities on a Continental scale'. For this he blamed the War Office, which prepared 'not for the next war, but for the last one or the last one but one'.[1] Jealously guarding its control over the production of guns and granting contracts to only a few private firms, Britain's national arsenals remained as obsolete, unprogressive and reactionary as they had in 1854. The country's leading armaments companies – Armstrong, Whitworth and Vickers – were excluded from the list of suppliers to the army. These great enterprises depended on the naval estimates and the Admiralty for their existence, and they had risen to their vast size after 1885 when the Admiralty acquired the right to buy guns from private firms.

We have also seen how Woolwich had thwarted government attempts to control it and avoid being pushed aside by Elswick, until the government was obliged to impose reforms on the War Office and brought in private engineers and gun makers to meet the crisis of the Crimean conflict. This process became a theme in British armaments history. The government remained completely unpredictable as they reacted only when a crisis was upon it, and deserted its private arms engineers when the crisis ended (usually thanks to those private engineers). The same broad features are evident in the post-1914 relationships between the government and Britain's armaments companies, and especially its oldest and most loyal one, which Sir William Armstrong had begun so long ago and placed at the service of his country. But the government always remained reluctant to commit the resources, even modest orders, to keep the firm afloat until the next great crisis fell upon them.

In 1914 the Sir William G. Armstrong Whitworth Company had grown into one of Britain's largest industrial enterprises, with a world-wide reputation for the highest-quality guns and mountings. The Armstrong and Elswick names had symbolised first rate innovative engineering since 1859, building a reputation which the company would continue to deserve both throughout the war and in the years of crisis and decline of the inter-war period. But the war was long and the demands put upon the company by the government left it more vulnerable to external forces than ever before, forces which now worked against the company having favoured it in 1863, when the British government withdrew its support once a crisis had passed. The structural changes wrought by the war are so great that the history of the company seems disconnected from its prewar past. In other words, 1914 is an appropriate point at which to bring the present work to a close because to go beyond 1914 would be to represent more of a break than a continuation in the history of the company, and to enter a period for which a separate work would be more appropriate.

The main point of this epilogue is to argue that the company's decline and fall in the inter-war period was a consequence of circumstances beyond its control: government decisions destroyed it armaments business in the 1920s and the Great Depression crushed its diversification into locomotives in the 1930s.

II

When the First World War broke out, Elswick was in the midst of passing out of the hands of its founders. Armstrong had died in 1900, Stuart Rendel died in 1913 and Noble's health was declining rapidly. His death in 1915 brought the long heroic phase of the company's history to a close. Control of the firm passed to the new generation under the chairmanship of John Meade Falkner. At first glance Falkner seems an unlikely candidate to run the vast armaments empire that Armstrong, Whitworth had become. He was not an engineer, a scientist or a businessman either by training or by inclination, but Oxford-educated in history, the classics and theology, who landed the job as tutor to Noble's children. Noble brought him into the company as his personal secretary, from which he worked his way up to company secretary, to director and vice-chairman. When he succeeded Noble in October 1915, Falkner realised that it would be absurd for him to pretend to lead Elswick in what he called the 'public' tradition of Sir William Armstrong and Sir Andrew Noble. Those men were world-famous figures of great scientific and engineering achievements. Yet Falkner did not equate his obscurity with Elswick, and recognised that other great companies were also in the hands of men who were not great themselves.[2] But the new men proved effective managers of Elswick. Indeed, one should say they more than fulfilled their promise, given the wartime crisis which tested their managerial and entrepreneurial talents in a way Armstrong and Noble never confronted. The new managers were relieved of seeking orders from foreign governments and the complicated political and marketing aspects which that entailed. The expansion of production far faster and further than anything in the past, and technological innovation to solve specific battlefield problems, consumed the time and energies of the company.

The demand for artillery and shells rose sharply in the autumn of 1914 and government orders flooded Elswick. In late August, the company was building 78 18-pounders for the government; in early October, the government increased that order to 460, to be delivered by August 1915. British generals soon discovered that they desperately needed heavy artillery (60-pounders) as well, and the main source of both these guns was Elswick. The company scrambled to erect new plant at Elswick and extended Whitworth's old Manchester plant at Openshaw. Later, Elswick also designed the guns and gun mountings for the first tanks after Colonel Swinton, the man largely responsible for the new weapon, discussed his idea with Admiral Ottley, one of Rendel's new recruits to the company. Elswick also remained the predominant source of naval guns and gun mountings. The fact that Elswick manufactured a greater variety of large guns than any other company increased its importance to the war effort.[3] During the war, Elswick manufactured 13 000 guns and 12 000 gun mountings, 14 500 000 shells, 18 million fuses and 21 million cartridge cases. The demand for warships tested the managerial and engineering capacities at Elswick, whose naval yards built more than 40 warships within the four-year period, including a 27 500-ton battleship, a 41 000-ton battle cruiser, a 10 000-ton aircraft carrier, 19 cruisers and 12 submarines. Elswick produced the naval guns for nearly 600 warships, built over 1000 aeroplanes and, towards the end of the war, 100 of the first tanks.[4]

Neither the government nor the private manufacturers were organised to accommodate the demand, nor did they recognise the complexity of mobilising both industry and the military for a large war against another powerful industrial nation. The recruitment of young patriotic engineers into the army, for example, depleted the ranks of industrial workers needed to produce the guns to fight the enemy. This loss of skilled workers from the armament companies was exacerbated by the unnecessary funnelling of skilled engineers into the unskilled work of producing shells. It was here that the system broke down. The infamous 'shell scandal' was the result of both poor industrial organisation and a false assumption about what kind of war was confronted. In August 1914, for example, the government ordered 162 000 18-pounder shrapnel shells from Elswick but doubled that order only two weeks later. By October, the government was demanding 400 000 shells and pressing for production to speed up from 35 000 to 55 000 per week.

Shell production, already approaching the limits of its capacity, was thrown into chaos when it was discovered that, however effective shrapnel shells had been against the Boers on the veldt of South Africa, they were inadequate against German troops in trenches. When the War Office suddenly ordered high-explosive shells from Elswick, the company's productive schedule, organised for the manufacture of shrapnel, was thrown into confusion. In May 1915, only a third of the six million shells due for delivery were finished and, when this shortfall halted a British counter-offensive at Ypres, the 'shell scandal' erupted. The irrational diversion of highly-skilled Elswick workers from their work on guns, mountings, ships and aeroplanes to the simple task of making shells – a scandal within the larger scandal of a shortage of shells – was a consequence brought about entirely by the government's failure to plan or estimate the number of shells they might require.

The shell shortage, the inability to break the stalemate in France and the appalling losses coincided with the failure of an Anglo-French fleet to break through the Dardanelles to Constantinople, knock Turkey out of the war, entice Italy and Greece to enter the war on the allied side, and open a supply line to the Russians, whose successes against Austria were faltering for lack of arms and ammunition. The campaign in Mesopotamia was also bogged down in late 1915. The military failures of 1915 produced a political crisis in Britain reminiscent of (but far more serious than) the crisis of 1854. The reaction was also similar (but far more radical). The ineffective wartime Prime Minister H.H. Asquith relinquished control to a Cabinet war committee, and a new Ministry of Munitions under the dynamic Lloyd George began to mobilise industry for war. The failure of the Somme offensive in 1916, with its unprecedented losses for little gain and nothing of strategic importance precipitated the replacement of the passive and detached Asquith, who left the war to generals and industry to businessmen, by the energetic, 'hands-on' mobiliser Lloyd George.

Lloyd George brought a sense of urgency and action to the war effort and expanded the military–industrial–political complex. He brought businessmen into the government as administrators and took a direct interest in all areas of the war effort, from strategic and tactical plans, such as the convoy system, to new weaponry, especially the tank. He devoted himself to overcoming the labour shortage, obtaining union agreement for the employment of unskilled workers and developing a close working relationship with industrial leaders upon whom the war effort had come to depend (to the chagrin of Britain's military leaders). Under the pressure of modern industrial war, Lloyd George loosened his commitment to Liberal *laissez-faire* ideology and moved towards greater government control, through negotiation with businessmen and unions, over the British economy. Industrialists, who had the technical and administrative knowledge to manage large establishments, were recruited to the Ministry of Munitions. As Sir William Armstrong and George Rendel had done many years previously, Elswick men joined the government in official capacities. Glyn West, for example, who became chairman of Elswick in 1921, was appointed Deputy Director General of Gun Ammunition.[5] Lord Sydenham resigned his seat on the board in 1916 'in order to undertake important Government work'.[6]

Meade Falkner, speaking as chairman to Elswick shareholders in 1915, proclaimed that 'the old order changeth' and that, 'instead of private enterprise, we have become a Government arsenal under Government control'.[7] However, the government did not nationalise the armaments firms, and prices and profits were negotiated rather than proclaimed by fiat.[8] Nevertheless, the Ministry of Munitions and the excess profits tax forced Elswick (and the other armaments firms) to expand beyond what a rational business strategy would have aimed towards and it was at this point that the company should have demanded a government guarantee about postwar business. The old danger remained of being abandoned by the government once the military–political crisis was resolved. When this occurred in 1863, the company had the opportunity to enter the foreign world markets. That option was not present after 1918. However, profits were very high during the war, and in 1919. Net profits for the four-year period 1916 to 1919 were £4 million on a share capital of £11 million.[9] The company was in strong financial shape when the war ended, but the period of adjustment to peacetime production would be full of uncertainties. New civilian investments must be found but, in the end, the fate of Britain's private armaments companies – Armstrong Whitworth, Vickers, Beardmore and all the others – would be determined by the foreign policies and diplomatic agreements of the British Cabinet.

III

The political, diplomatic, strategic, financial and international context of the armaments business in 1919 was completely altered by the war. The pre-war claim that great armaments ensured peace was transformed into its opposite: that, in Sir Edward Grey's famous phrase, 'great armaments lead inevitably to war'.[10] Many in Britain had hoped and assumed that they could return to the world of 1914, but they did not mean a return to the era of belligerent international relations, arms races and great navies. By 1925, the company faced an impossible situation: the high wartime demand for armaments had ended in 1918; the naval armaments treaty clauses in the Washington Conference of 1922 had eliminated the international market for naval armaments;[11] the Treasury had halted the naval and air force rearmament programme of the early 1920s, aimed at maintaining control over the empire and an independent diplomatic role in world affairs; the Ten Year Rule had laid a moratorium on naval armaments production; and the bomber had replaced the battleship as the greatest threat to Britain.[12] Elswick's managers were well aware that the high demand for warships was temporary and as early as 1916 they began planning postwar adjustments into civil engineering and new areas such as aeroplanes and engines, locomotives, motor cars and tools, or any engineering business that might fit Elswick's facilities. However, the company was overwhelmed by the degree to which postwar external circumstances changed and the speed with which they needed to find places to invest several million pounds.

As the war drew to an end in 1918, the board of directors made the strategic decision to enter the railroad locomotive business. In keeping with the Elswick tradition of producing high-quality guns and mountings, an experienced American locomotive engineer was brought over to build the most modern, efficient and complete locomotive manufacturing plant. The Scotswood Locomotive Plant was an immediate success in the Elswick tradition of selling into the global markets. No doubt using their international reputation and innumerable contacts, orders were won to supply locomotives for railroad companies in Britain, Buenos Aires, India, Belgium, Australia, Romania and elsewhere.[13] It seemed that the post-1863 recovery was to be repeated. Locomotives were an ideal substitute for the company, in that they had experience with the advanced engineering and international marketing which involved governments. In many ways manufacturing and selling locomotives was a rational diversification for the company to wean itself off total dependence on armaments. If the world economic recovery continued, the company would successfully transform itself into a general engineering company with locomotive production for global markets at its core.

The move into locomotives was the right decision, based on the rational belief in the growth potential of the postwar global economy and that railroads would be an important capital goods sector of that recovery. The economic signs in 1919–20 all pointed to recovery and engineering companies everywhere committed their capital to participate in the new industrial age. Scotswood was an efficient, well-equipped plant with a sound business plan, and its locomotives sold well throughout the inter-war period despite the high capital costs involved, aggressive competition and the economic depression of the 1930s. It is a great irony that the Scotswood Locomotive Plant was bought by the

British Admiralty in 1937 and turned over to Vickers-Armstrong to be refurbished as an armaments works.

Locomotives were but one part of the company's strategy of diversifying into civilian engineering. New businesses had to be found for a company whose productive capacity had doubled during the war. Worn out by the war, Falkner resigned in 1920 and was succeeded by Sir Glynn West from the Ministry of Munitions. The company invested in a variety of 'peace industries' in 1920: Siddeley Deasy Motor Car Co., Pearson & Knowles Coal & Iron Co., Armstrongs & Main, Ltd., Crompton & Co. and numerous others, including Charles Walmsley & Co., a well-known manufacturer of paper-making machinery.[14] But the directors admitted that 'until the general conditions of manufacture become more stable it is difficult to make any forecast of the future'.[15] General conditions did not improve. Indeed, Britain's economic situation deteriorated, especially in the iron, steel and ship-building industries. Unlike the 1880s, the political and international context prompted a reduction in government naval orders which came in 1921, just as merchant shippers were cancelling orders to Elswick's merchant ship-building yard, squeezing Elswick severely in the midst of their diversification and expansion programme. Strikes exacerbated the company's weakening position. In 1922, the company reported that 'the depression in trade continues and the difficulties have been accentuated by the further industrial disputes this year lasting more than three months and affecting all branches of the Company's business'. A strike by boiler makers in 1923 lasted many months and caused 'considerable trouble in the Shipyards' and 'seriously affected the profits of the Company'.[16]

The diversification strategy was sound but risky, in that the necessary capital costs of were raised through debt financing. In 1922, the company launched a major investment programme in a pulp and paper mill in Newfoundland, with vast forests of spruce trees, a softwood which could be mechanically turned into pulp and paper for the American mass market for cheap newspapers. The pulp and paper mill was envisioned as a source of demand for electrical and engineering products made by other departments in the Armstrong Whitworth company: engines and heavy engineering equipment from Elswick and Openshaw, pulp and paper-making machinery from Walmsley and merchant ships from Elswick.[17] It was a strategic plan of vision designed to rebuild Armstrong Whitworth into an integrated, self-developing and interdependent system of production. It was bold and very risky. Between 1922 and 1926, debt increased from £5 million to £14 million, while share capital remained at £11 million.[18] The strategy failed because it rested on the assumption that the world economic cycle would once again turn upwards. The increased risk came from increasing the company's dependence on external forces which only governments had any hope of controlling. If it had succeeded, the company would have gone down in history as carrying out one of the greatest entrepreneurial feats of the age. Their plans were thwarted by an accumulation of external political and economic factors which brought the postwar boom to an end: the speedy lifting of government wartime controls and the reprivatising of the coal and railroad industries, the rejection of electrifying and rationalising British rail transport, the sudden reduction in ship building following the Washington Conference in 1922, the failure of governments to develop a rational reparations settlement, the recession of 1924, and the return to the gold standard in 1925. Yet, despite these difficult circumstances, which exacerbated the company's precarious financial situation in the early 1920s, expansion through acquisitions was the only rational choice for a company of the size of Armstrong Whitworth if it was to have any chance of survival.

Disaster came from Newfoundland where over £5 million was invested in the Newfoundland Power & Paper Co., of which £3 million was raised through mortgage debentures. The scheme required the building of a town site at Corner Brook, along with the warehouses, docks, ships and machinery needed to operate a pulp and paper company. It was a risky venture by a financially weak company. As the leading business periodical suggested, the project had to produce a return in a relatively short period of time if Armstrong shareholders were to remain confident.[19] In 1925, the company directors reported further deterioration in the company's position. The trade depression of 1924–5 was deeper than they had anticipated and they pointed to lower wages in Europe and fierce competition 'both in the home and other markets'.[20] The situation became very serious in 1926. The strenuous efforts over the previous eight years to replace armaments with civilian products were 'not giving the results which were expected'. The company was forced to admit that 'the original estimates of profit in connection with important contracts will not be realized and that, instead, heavy losses due to the above mentioned and other causes will result'.[21] Losses for 1925 were £900 000. In 1926, the year of the great coal strike, the company reported a loss of nearly £700 000.[22]

Assets, including the Newfoundland paper company, were sold to cover the growing losses, but by 1926 the company was facing bankruptcy. The Pozzuoli works were already in liquidation, with a loss of £13 000 000 and the Pearson and Knowles group of companies was sold at a loss of £1 500 000. Armstrong Siddeley and Walmsley were also sold.[23]

The Bank of England had been a prominent participant in the

Newfoundland project and wrote off over £5 million.[24] The crisis required drastic action and, in 1927, Sir W.G. Armstrong, Whitworth and Company – Elswick – was amalgamated with Vickers, its great prewar rival for naval armaments and which was also suffering the hardships of peace. Two companies were formed: the English Steel Corporation and Vickers-Armstrong, which built naval ships and armaments. A new company was formed in 1929, the Armstrong Whitworth Securities Company Limited, which had a share capital of less than £5 million and a debt of £1 500 000. Its assets included £900 000 in Vickers-Armstrong stock and nearly £1 million in cash and government securities. Armstrong Whitworth Securities owned three operating companies: Sir W.G. Armstrong, Whitworth & Company (Engineers) Limited; Sir W.G. Armstrong, Whitworth & Company (Shipbuilders); and Sir W.G. Armstrong, Whitworth and Company (Ironfounders) Limited.[25]

The new company made a profit of £12 000 in 1929 and £57 000 in 1930, looked upon by the chairman as 'very satisfactory'. The company had begun to build diesel engines for locomotives and, in a joint venture with a Swiss company, diesel lorries 'for the home market and for the Empire overseas'.[26] The company struggled to survive in the 1930s but in 1936 was forced to sell its Vickers-Armstrong shares; in 1937, the locomotive plant was also sold.[27] There was nothing left. The company was dismantled because, even after 1933, public opinion held that a primary cause of the war had been armaments, a view that suited the financial constraints of the Treasury. The British government's neglect of British armaments in the inter-war period, when there was no substitute global armaments market for companies like Armstrong and Vickers, left British diplomacy and foreign policy hesitant, inconsistent and without the confidence of the Victorian and Edwardian era as the clouds gathered, bringing another storm of steel and a new generation of weapons of mass destruction.

Notes

1. David Lloyd George, *War Memoirs*, vol. 1: 1914–1915 (Boston: Little, Brown, 1933), p. 112.
2. Letter to Philip Noble, 1 Jan. 1916, on succeeding to his father's position, reproduced in Kenneth Warren, *Armstrongs of Elswick: Growth in Engineering and Armaments to the Merger with Vickers* (Basingstoke: Macmillan, 1989), p. 189. Warren reproduces many unedited letters which provide much insight into the personalities of the company directors and their views of each other.
3. Scott, *Vickers: A History* (London: Weidenfeld & Nicolson, 1962), pp. 99, 104, 107, 114–15.

4. Warren, *Armstrongs*, pp. 190–91.
5. Scott, *Vickers*, p. 104.
6. Sir W.G. Armstrong, Whitworth & Co., Ltd., *Twenty-First Report* (1916), T&W, Armstrong Archives.
7. Ibid., p. 126.
8. Ibid., pp. 127–30.
9. No annual reports were published for 1916, 1917 and 1918. Armstrong, Whitworth, *Twenty-Third Report* (1921) covers the four years 1916 to 1919.
10. Sir Edward Grey, *Twenty-Five Years 1892–1916*, vol. 1 (New York: Frederick Stokes, 1925), p. 89.
11. Amongst other things, the treaty called for a 10-year period during which no new capital ships (those of 10000 tons and 8-inch or larger guns) were to be built. Further restrictions were added at the London Naval Conference of 1930 (which Germany did not sign). The Anglo-German Naval Agreement of 1935 granted Germany the right to expand her navy.
12. David French, *The British Way in Warfare 1688–2000* (London: Unwin Hyman, 1990), pp. 180–84.
13. Warren, *Armstrongs*, pp. 205–6.
14. *Twenty-Fourth Report* (1922).
15. Armstrong, Whitworth, *Twenty-Second Report* (1920).
16. *Twenty-Fifth Report* (1923), *Twenty-Sixth Report* (1924).
17. Warren, *Armstrongs*, pp. 215, 217.
18. *The Economist*, 18 Dec. 1926, p. 1059.
19. *The Economist*, 20 Dec. 1924, p. 1021.
20. *Twenty-Seventh Report* (1925).
21. *Twenty-Eighth Report* (1926).
22. *Twenty-Eighth Report* (with balance sheet for 1925) and *Twenty-Ninth Report* (with balance sheet for 1926).
23. *The Economist*, 11 June 1927, p. 1228; 16 Feb. 1929, pp. 347, 373.
24. *The Economist*, 16 Feb. 1929, pp. 347, 373.
25. *The Economist*, 'Steel Reorganization: A First Step', 22 Dec. 1928, pp. 1149–50 and 'The Armstrong Whitworth Reconstruction', 16 Feb. 1929, pp. 347–8. See also Armstrong Whitworth Securities, *Thirty-Second Report* (1930), *Thirty-Third Report* (1931).
26. *The Economist*, 20 June 1931, p. 1344.
27. *The Economist*, 7 Aug. 1937.

Bibliography

Manuscript sources

Admiralty Papers, Public Record Office, London.

Armstrong, Whitworth Records, Tyne & Wyre Public Archives, Newcastle. This is a collection of a variety of material relating to the business of the company from 1882: minute books, pamphlets and a vast amount of miscellaneous engineering material such as technical documents and drawings. This collection was of little use for this work.

Campbell Papers, School of Oriental and African Studies, London.

D'Eyncourt Papers, National Maritime Museum, London.

Foreign Office Papers, Public Record Office, London.

Rendel Letters and Papers, National Library of Wales.

Rendel Papers, privately held by Miss Rosemary Rendel, London. These are letters from Hamilton Rendel to Stuart Rendel.

Rendel Papers, Tyne & Wyre Public Archives, Newcastle. This collection of 10 000 items (mostly letters to Stuart Rendel) constitutes a large and valuable archival source of information about many aspects of the company. It begins in 1863 and ends with Rendel's death in 1913.

Vickers Papers, Vickers House, London. This collection includes various items related to Armstrong, Whitworth after about 1880: letters, books, pamphlets.

War Office Papers, Public Record Office London.

Printed primary sources

Parliamentary Papers

The reports are listed in chronological order.

Reports from the Select Committee on the Army Before Sebastopol (Roebuck Committee), IX (1854–5).

Report of a Committee of the Treasury to Inquire into the Navy Estimates, from 1852 to 1858, and into the Comparative States of the Navies of England and France (Derby Committee), XIV, Part 1 (1859).

Report of the Royal Commission on the Defences of the United Kingdom, XXIII (1860).

Return of Weapons of War, Accounts and Papers, vol. 3, XLI (1860).

Select Committee on Military Organization, VII (1860).

Report on the Board of Admiralty, V (1861).
Report of the Defence Commissioners on Spithead Forts, XXVII (1862).
Report of the Ordnance Select Committee, VI (1862).
Report from the Select Committee on Ordnance into the Expenditure Incurred Since 1858 on Various Natures of Improved Ordnance, XI (1863). This 800-page report is a mine of information about Armstrong's guns and Elswick sales to the British government. The following four statements from its appendices are of particular value.
Statement Showing the Amount and Rate Paid for Certain Guns to the Elswick Ordnance Company Compared with Original Rates, XI (1863).
Statement Showing the Cost of Certain Guns and Projectiles Obtained from the Elswick Ordnance Company, XI (1863).
Statement Showing the Payments to the Elswick Ordnance Company, XI (1863).
Statement Showing Total Amount Expended in the Manufacture of Guns and Projectiles on Sir William Armstrong's Principle, XI (1863).
Return of the Number and Sizes of Armstrong Guns Supplied to the Government from Elswick and from Woolwich, XXXII (1863).
Sir William Armstrong: Correspondence, XXXII (1863).
Armstrong Guns: Returns of Expenses Incurred on all Classes of Armstrong Guns Showing Details of Alterations and Changes, and the Present Cost of Breech-loading and other Armstrong Guns, XXXII (1865).
Report of Admiral Kuper in reference to the Armstrong Guns in the Action of Simonosaki, XXXII (1865).
Report of Admiralty Committee on Turret Ships, XLVII (1866).
Report of the Special Committee on Armstrong and Whitworth Guns, XLII (1866).
Papers Relating to the Dismissal of Colonel Boxer, XLII (1870).
Report of the Committee Appointed by the Admiralty to Examine the Designs Upon Which Ships of War Have Recently Been Constructed, XIV (1872).
Recommendations of the Ordnance Committee (with Special Associated Members) As to the Construction of Ordnance, XIV (1884–5).
Report of the Committee Appointed to Inquire into the Conditions Under Which Contracts are Invited for the Building or Repairing of Ships, Including their Engines, For Her Majesty's Navy, and into The Mode in Which Repairs and Refits of Ships are Effected in Her Majesty's Dockyards (Ravensworth Committee), XIV (1884–5).
Report of the Committee Appointed by the Admiralty to Inquire into the System of Purchase and Contract in the Navy (Forwood Committee), XVI (1887).

Report of the Committee on the Organization and Administration of the Manufacturing Departments of the Army (Morley Committee), XIV (1887).

Reports from the Select Committee on Army and Navy Estimates, VIII (1887). There are five of these reports.

Royal Commission Appointed to Inquire into the System Under Which Patterns of Warlike Stores are adopted and the Stores Obtained and Passed for Her Majesty's Service, XV (1887).

Reports of the Select Committee on Naval Estimates, Accounts and Papers, XIII (1888).

Return Describing the Position and Capacity of each Furnace in the Royal Arsenal, Woolwich, LXVII (1888).

Return of Letter from Certain Steel Manufacturers of Sheffield to the Chairman of the Select Committee on Army Estimates Together with His Reply, LXXXI (1888).

Report of the Committee on War Office Organization (Dawkins Committee), XL (1901).

Report of the Committee Appointed to Inquire into the Arrears of Shipbuilding, LXI (1902).

Report of the Conference Appointed to Examine the Shops and Machinery at Woolwich Arsenal in order to Consider Whether any article now not made in the Ordnance Factories can Appropriately be made there with this machinery, XLIX (1907).

Report of the Government Factory and Workshops Committee (Murray Committee), X (1907).

Other

Hansard (1854–1914).

Printed secondary sources

Reference Works

Brassey, Thomas, *The Naval Annual*, New York, Praeger.

Debrett's House of Commons and the Judiciary Bench

Dictionary of Business Biography

A Dictionary of Edwardian Biography: Northumberland, Edinburgh: Peter Bell, 1985.

Dictionary of Military Terms: A Guide to the Language of Warfare and Military Institutions, compiled by Trevor Dupuy, Curt Johnson and Grace Hayes, New York: The H.W. Wilson Company, 1986.

Dictionary of National Biography
Dictionary of Scientific Biography
Encyclopaedia Britannica, 9th edn (1890); 10th edn (1902); 11th edn (1910); 15th edn (1985, first published 1974).
Hartman, Tom, *The Guinness Book of Ships & Shipping: Facts & Feats*, London: Guinness Superlatives Limited, 1983.
Kemp, Peter, ed., *The Oxford Companion to Ships & The Sea*, London: Oxford University Press, 1976.
Layton, C.W.T., *Dictionary of Nautical Words and Terms*, 2nd rev. edn by Peter Clissold, Glasgow: Brown, Son & Ferguson, Ltd, 1982; first published 1955.
The National Encyclopedia, New York: P.F. Collier, 1936.
The New International Encyclopaedia, New York: Dodd, Mead & Comp., 1903.
Oxford English Dictionary
Pike, W.T., ed., 'Contemporary Biographies', *Northumberland at the Opening of the Twentieth Century*, Brighton: W.T. Pike & Co., 1905.
Quick, John, *Dictionary of Weapons and Military Terms*, New York: McGraw-Hill Book Company, 1973.
Report of Patent, Design, Trade Mark, & Other Cases, vols 20 (1903), 22 (1905), 23 (1906), 24 (1907).
Stenton, Michael, *Who's Who of British Members of Parliament: A Biographical Dictionary of the House of Commons*, 2 vols, Hassock, Sussex: Harvester Press.
The Statist
Tarassuk, Leonid and Blair, Claude, eds, *The Complete Encyclopaedia of Arms & Weapons*, trans. of Italian edn (1979) Sylvia Mulcahy, London: B.T. Batsford Ltd, 1982.
Who's Who in History
Who Was Who
World Who's Who In Science, Chicago: Marquis Who's Who, 1968.

Periodicals

Arms and Explosives
The Army and Navy Magazine
Blackwood's Magazine
Colburns United Service Magazine
Economist
Edinburgh Review
Engineer
Fraser's Magazine
The Graphic

The Illustrated London News
London Review of Politics, Society, Literature, Art and Science
Mechanics Magazine
Quarterly Review
The Times: Armstrong and other engineers debated, challenged each other and the government, and otherwise communicated amongst themselves and to the British public in their many letters to the editor of this paper. The paper published many articles about Armstrong and his guns, and printed his public speeches.
Transactions, Royal Institute of Naval Architects.

Books and Articles

Adye, Sir John, *Recollections of a Military Life*, privately published, 1895.

Appleyard, Rollo, *Charles Parsons: His Life and Work*, London, 1933.

Armstrong, Sir William, 'Our National Defences', *Colburn's United Service Magazine* 158 (1882).

Armytage, W.H.G., 'Sheffield and the Crimean War: Politics and Industry 1852–1857', *History Today* 5 (1955).

'Artifex' and 'Opifex' [pseud], *The Causes of Decay in a British Industry*, London: Longmans, 1907.

Atlay, James B., *The Victorian Chancellors*, London: Smith, Elder, 1906–8.

Ballard, Admiral George A., *The Black Battlefleet*, Lymington, Hampshire: Nautical Publishing Company Limited, 1980.

——, *The Influence of the Sea on the Political History of Japan*, London: John Murray, 1921.

Barnaby, Sir Nathaniel, 'On Mechanical Invention in its Relation to the Improvement of Naval Architecture', *Transactions*, vol. 1, London: Royal Institute of Naval Architects, 1860.

——, 'On Ships of War', *Transactions of the Institute of Naval Architects*, vol. 17, 1876.

——, 'Shipbuilding', *Encyclopaedia Britannica*, 9th edn, 1890, vol. 21, pp. 809–26.

——, *Naval Developments of the Century*, London: Linscott Publishing, 1904.

Barnes, Leslie W.C.S., *Canada's Guns: An Illustrated History of Artillery*, Ottawa: Canadian War Museum, 1979.

Bartlett, C.J., 'The Mid-Victorian Reappraisal of Naval Policy', in K. Bourne and D.C. Watt (eds), *Studies in International History*, London: Longmans, 1967, pp. 189–208.

Bastable, Marshall J., 'From Breechloaders to Monster Guns: Sir William Armstrong and the Invention of Modern Artillery, 1854–1880', *Technology and Culture* 33 (1992), 213–47.

Batzel, Victor M., 'Legal Monopoly in Liberal England: The Patent Controversy in the Mid-Nineteenth Century', *Business History* 22 (1980), 189–202.

Baxter, James P., *The Introduction of the Ironclad Warship*, Cambridge, Mass.: Harvard University Press, 1933.

Beckett, I.F.W. and Gooch, John, *Politicians and Defence: Studies in the Formation of British Defence Policy 1845–1970*, Manchester: Manchester University Press, 1981.

Bell, Herbert C., *Lord Palmerston*, 2 vols, London: Longman's, 1936.

Best, Geoffrey, 'The Militarization of European Society, 1870–1914', in John R. Gillis (ed.), *The Militarization of the Western World*, New Brunswick and London: Rutgers University Press, 1989, pp. 13–29.

Bijker, Wiebe, Hughes, Thomas and Pinch, Trevor (eds), *The Social Construction of Technological Systems: New Directions in the Sociology and History of Technology*, Cambridge, Mass.: MIT Press, 1987.

Birch, Alan, *The Economic History of the British Iron and Steel Industry 1784–1879: Essays in Industrial and Economic History with Special Reference to the Development of Technology*, London: Frank Cass, 1967.

Bishop, Farnham, *The Story of the Submarine*, New York: Century, 1916.

Blakely, Mrs T., *Case for 'Justice'*, London: n.p., 1902.

——, *Blakely and the Armstrong Guns: A Statement*, London: n.p., 1898.

Blakely, Theophilus, *A Letter from Captain Blakely, H.P., Royal Artillery, to the Secretary of State for War Claiming the Original Invention of an Indispensable Feature of The Armstrong Gun*, London: J. Ridgway, 1859.

Block, I.S., *The Future of War in its Technical, Economic, and Political Relations*, Boston: Ginn, 1899.

Blumenthal, Harvey, 'W.T. Stead's Role in Shaping Official Policy: The Navy Campaign of 1884', Ph.D thesis, The George Washington University, 1984.

Borton, Hugh, *Japan's Modern Century*, New York: The Ronald Press, 1955.

Boulger, Demetrius C., *The Life of Sir Halliday Macartney*, London: J. Lane, 1908.

Bourne, Kenneth, *The Foreign Policy of Victorian England 1830–1902*, Oxford: Clarendon Press, 1970.

Bousel, Patrice, *Leonardo da Vinci* (Secaucus, N.J.: Chartwell Books, 1986).

Briggs, Sir John Henry, *Naval Administration, 1827 to 1892*, London: S. Low, Marston, 1897.

Brockway, Fenner, *The Bloody Traffic*, London: Victor Gollancz, 1933.

Brodie, Bernard, *Sea Power in the Machine Age*, 2nd edn, Princeton, N.J.: Princeton University Press, 1943.

——, 'Technological Change, Strategic Doctrine, and Political Outcomes', in Klaus Knorr (ed.), *Historical Dimensions of National Security Problems*, Lawrence, Kansas: University of Kansas Press, 1976, pp. 263–306.

Browne, Douglas G., *The Floating Bulwark: The Story of the Fighting Ship 1514–1942*, London: Cassell, 1963.

Bruce, Maurice, *The Shaping of the Modern World 1870–1939*, vol. I, *Ends and Beginnings: The World to 1914*, London: Hutchinson, 1958.

Brunel, Isambard, *The Life of Isambard Kingdom Brunel*, London, 1870, reprinted with introduction by L.T.C. Rolt, London: David & Charles, 1971.

Buckle, George E., *The Letters of Queen Victoria*, 2nd series, 2 vols, London: J. Murray, 1926.

Burn, Duncan, *The Economic History of Steel Making 1867–1939*, Cambridge: University of Cambridge Press, 1961.

Burnham, T.H. and Hoskins, G.O., *Iron and Steel in Britain 1870–1930: A Comparative Study of the Causes which Limited the Economic Development of the British Iron and Steel Industry between the Years 1870 and 1930*, London: Allen and Unwin, 1943.

Busk, Hans, *The Navies of the World*, London: Routledge, Warnes, Routledge, 1859; reprinted by the Naval Institute, 1973.

Cadogan, Peter, *Early Radical Newcastle*, Consett: Sagittarius Press, 1975.

Cameron, Meribeth, Mahoney, Thomas and McReynolds, George, *China, Japan and the Powers: A History of the Modern Far East*, 2nd edn, New York: The Ronald Press Company, 1960; first published 1952.

Cardwell, Donald, *The Fontana History of Technology*, London: Fontana Press, 1994.

Carr, J.C. and Taplin, W., assisted by Wright, A., *History of British Steel Industry*, Cambridge, Mass.: Harvard University Press, 1962.

Cavendish, Francis W.H., *Society, Politics and Diplomacy 1820–1864: passages from the journal of Francis W.H. Cavendish*, London: T.F. Unwin, 1913.

Cecil, Lady Gwendolen, *Life of Robert Marquis of Salisbury*, vol. 4, London: Hodder and Stoughton, 1932.

Chandler, Alfred D., 'The Development of Modern Management Structure in the U.S. and U.K.', L. Hannah (ed.), *Management Strategy and Business Development: An Historical and Comparative Study*, London: Macmillan, 1976.

——, *The Visible Hand: The Managerial Revolution in American Business*, Cambridge, Mass., The Belnap Press of Harvard University Press, 1977.

——, *Scale and Scope: The Dynamics of Industrial Capitalism*, Cambridge, Mass.: Harvard University Press, 1990.

Chatfield, Lord, *The Navy and Defence: The Autobiography of Admiral of the Fleet Lord Chatfield*, London: Heinemann, 1942.

Chesneaux, Jean, Bastid, Marianne and Bergère, Marie-Claire, *China From the Opium Wars to the 1911 Revolution*, trans. Anne Destenay, New York: Random House, 1976.

Childers, Lt Col. Spencer, *The Life and Correspondence of the Right Hon. C.E. Childers, 1827–1896*, vol. 1, London: Murray, 1901.

Church, William C., *The Life of John Ericsson*, New York: Charles Scribner's Sons, 1911; first published 1890.

Cipolla, Carlo, *Guns and Sails in the Early Phase of European Expansion 1400–1700*, London: Collins, 1965.

Clapham, J.H., *An Economic History of Modern Britain: Free Trade and Steel, 1850–1886*, Cambridge: Cambridge University Press, 1932.

Clowes, William L., *The Royal Navy: A History From the Earliest Times to the Present*, 7 vols, London: Sampson Low, Marston and Company, 1897–1903.

Cobden, Richard, *The Three Panics*, London, 1862.

Cobden Club, *The Burden of Armaments: A Plea for Retrenchment*, London: T.F. Unwin, 1905.

Collier, Basil, *Arms and the Men: The Arms Trade and Governments*, London: Hamish Hamilton, 1980.

Comparato, Frank E., *Age of Great Guns: Cannon Kings and Cannoneers Who Forged the Firepower of Artillery*, Harrisburg, Pa.: Stackpole, 1965.

Conacher, J.B., 'Party Politics in the Age of Palmerston', in Philip Appleman, William Madden and Michael Wolff (eds), *1859: Entering An Age of Crisis*, Bloomington: Indiana University Press, 1959, pp. 163–80.

Congreve, William, *An Elementary Treatise on the Mounting of Naval Guns*, London, 1811; reprinted by the Museum Restoration Service, Ottawa, 1970.

Conway's All the World's Fighting Ships 1860–1905, London: Conway Maritime Press, 1979.

Cook, Edward, *The Life of Florence Nightingale*, 2 vols, New York: Macmillan, 1942.

Craig, F.W.S., *British Parliamentary Election Results 1832–1885*, London: Macmillan, 1977.

Craig, Gordon, *Germany 1866–1945*, New York: Oxford University Press, 1978.

Creveld, Martin van, *Technology and War: From 2000 B.C. to the Present*, London: The Free Press, 1989.

Crowl, Philip A., 'Alfred Thayer Mahan: The Naval Historian', in Peter Paret (ed.), *Makers of Modern Strategy from Machiavelli to the Nucclear Age*, Princeton: Princeton University Press, 1986, pp. 444–77.

Cutliffe, Stephen H. and Post, Robert C. (eds), *In Context: History and the History of Technology: Essays in Honor of Melvin Kranzberg*, Bethleham, Pa.: Lehigh University Press, 1988.

Dasent, Arthur I., *John Thadeus Delane: His Life and Correspondence*, 2 vols, New York: Scribner's Sons, 1908.

Davenport-Hines, R.P.T., 'Introduction', in R.P.T. Davenport-Hines (ed.), *Markets and Bagmen: Studies in the History of Marketing and British Industrial Performance 1830–1939*, London: Gower, 1986, pp. 1–15.

——, 'The British Marketing of Armaments 1885–1935', in R.P.T. Davenport-Hines (ed.), *Markets and Bagmen: Studies in the History of Marketing and British Industrial Performance 1830–1939*, London: Gower, 1986.

——, 'Vickers As a Multinational Before 1945', in Geoffrey Jones (ed.), *British Multinationals: Origins, Management and Performance*, London: Gower, 1986, pp. 43–74.

Deacon, Richard, *A History of the British Secret Service*, New York: Tapinger, 1969.

Denholm, Anthony, *Lord Ripon 1827–1909: A Political Biography*, London: Croom Helm, 1982.

Derry, T. K. and Williams, T.I., *A Short History of Technology*, Oxford: Oxford University Press, 1960.

d'Ombrain, Nicholas, *War Machinery and High Policy: Defence Administration in Peacetime Britain 1902–1914*, Oxford: Oxford University Press, 1973.

Dougan, David, *The Great Gunmaker: The Story of Lord Armstrong*, Newcastle: Graham, 1971.

Douglas, Sir George B. and Dalhousie, Sir George (eds), *The Panmure Papers: Selections from the Correspondence of Fox Maule, 2nd Baron Panmure*, 2 vols, London: Hodder and Stoughton, 1908.

Dredge, James, *Modern French Artillery*, London: reproduced from 'Engineering', 1892.

Dreyer, E.L., *Early Ming China; A Political History, 1355–1435*, Stanford: Stanford University Press, 1982.

Eardley-Wilmot, Sydney M, *The Development of Navies During the Last Half Century*, London: Seeley, 1892.

——, *An Admiral's Memories*, London: Sampson Low, Marston, 1927.

Egerton, Mrs Fred, *Admiral of the Fleet Sir Geoffrey Phipps Hornby*, Edinburgh: W. Blackwood, 1896.

The Elswick Ordnance Company, n.p., 1912.

Emmerson, George, *John Scott Russell: A Great Victorian Engineer and Naval Architect*, London: John Murray, 1977.

Engelbrecht, H.C. and Hanighen, F.C., *Merchants of Death: A Study of the International Armament Industry*, London: George Rutledge, 1934.

Fairbairn, A.R., *Elswick*. n.d. Vickers Archive, no. 107.

Fairbairn, William, *Iron: Its History, Properties, and Processes of Manufacture*, Edinburgh: Adam and Charles Black, 1861.

——, *Treatise on Iron Shipbuilding: Its History and Progress*, London: Longmans, Green, and Co., 1865.

Fairbank, John K., Bruner, Katherine and Matheson, Elizabeth (eds), *The I.G. in Peking: Letters of Robert Hart, Chinese Maritime Customs 1868–1907*, 2 vols, Cambridge, Mass.: Harvard University Press, 1975.

Fairbank, John K., Reischauer, Edwin and Craig, Albert, *East Asia: Tradition and Transformation*, Boston: Houghton Mifflin Company, 1976.

Falk, Edwin A., *Togo and the Rise of Japanese Sea Power*, New York: Longmans, Green and Co., 1936.

Feuerwerker, Albert, 'Economic Trends in the Late Ch'ing Empire, 1870–1911', in John K. Fairbank and Kwang-ching Liu (eds), *The Cambridge History of China*, vol. 11, *Late Ch'ing, 1800–1911*, Part 2, Cambridge: Cambridge University Press, 1980, pp. 1–69.

Ffoulkes, Charles, *The Gun-Founders of England: With a List of English and Continental Gun-Founders from the XIV to the XIX Centuries*, Cambridge: Cambridge University Press, 1937.

Forbes, A., *A History of the Army Ordnance Services*, 2 vols, London: Medici Society, 1929.

Fortescue, J.W., *A History of the British Army*, vol. XIII, 1852–1870, London: Macmillan, 1930.

Fraser, Mary, *A Diplomatist's Wife in Many Lands*, vol. II, New York: Dodd, Mead, 1913.

French, David, *The British Way in Warfare 1688–2000*, London: Unwin Hyman, 1990.

Friedman, Norman, *Modern Warship: Design and Development*, Greenwich: Conway Maritime Press, 1979.

Fullom, Stephen W., *The Life of General Sir Howard Douglas*, London: John Murray, 1863.

Garbett, Captain H., *Naval Gunnery: A Description and History of the Fighting Equipment of a Man-of-War*, London: George Bell & Sons, 1897.

Gosses, F., *The Management of British Foreign Policy Before the First World War*, trans. E.C. Van Der Gaaf, Leiden: A.W. Sijthoff, 1948.

Graham, G.S., *The Politics of Naval Supremacy: Studies in British Maritime Ascendancy*, Cambridge: Cambridge University Press, 1965.

Grant-Duff, M.E., *Notes from a Diary*, 2 vols, New York: Dutton, 1905.

Great Britain, Committee of Imperial Defence, 3 vols, *Official History of the Russo-Japanese War*, vol. 1, London: H.M.S.O., 1910–20.

——, Ministry of Munitions, *History of the Ministry of Munitions*, 8 vols, London: H.M.S.O., 1922.

Grey, Sir Edward, *Twenty-Five Years 1892–1916*, vol. 1, New York: Frederick Stokes, 1925.

Guedalla, Philip, *Gladstone and Palmerston: Being the Correspondence of Lord Palmerston with Mr. Gladstone 1851–1865*, London: Victor Gollancz, 1928.

Guilmartin, John F., *Gunpowder and Galleys: Changing Technology and Mediterranean Warfare at Sea in the Sixteenth Century*, Cambridge: Cambridge University Press, 1974.

Haber, L.F., *The Chemical Industry During the Nineteenth Century: A Study of the Economic Aspect of Applied Chemistry in Europe and North America*, Oxford: Clarendon, 1958.

Hall, A.R., *Ballistics in the Seventeenth Century: A Study in the Relations of Science and War with Reference Principally to England*, New York: Harper & Row, 1969; first published by Cambridge University Press, 1952.

Hamilton, C.E., 'The Royal Navy, *La Royale*, and the Militarization of Naval Warfare, 1840–1870', *Journal of Strategic Studies* 6 (1983), pp. 182–212.

Hamilton, C.I., *Anglo-French Naval Rivalry 1840–1870*, Oxford: Clarendon Press, 1993.

Hamilton, George, *Parliamentary Reminiscences and Reflections 1886–1906*, 2 vols, London: Murray, 1922.

Hamilton, Admiral R. Vesey, *Naval Administration*, London: G. Bell and Sons, 1896.

Hamilton, W.M., 'The Nation and the Navy: Methods and Organization of British Navalist Propaganda 1889–1914', Ph.D thesis, London, 1977.

Hao, Yen-P'ing and Erh-min Wang, 'Changing Chinese Views of Western Relations, 1840–95', in John K. Fairbank and Kwang-ching Liu (eds), *The Cambridge History of China*, vol. 11, *Late Ch'ing, 1800–1911*, Part 2, Cambridge: Cambridge University Press, 1980, pp. 142–201.

Harcourt, H.W. Vernon, 'Our Naval and Military Establishments with Reference to the Dangers of Invasion', *Journal of the Royal United Service Institution* 16 (1872).

Harris, J.R., *The British Iron Industry 1700–1850*, London: Macmillan Education Limited, 1988.

Harvey, Frederick, *Instructions for the Management of Harvey's Sea Torpedo*, London: 1871.

Hawkins, C.J., *Theory of the Firm*, London: Macmillan, 1973.

Hazlett, James, Olmstead, Edwin and Parks, M., *Field Artillery Weapons of the Civil War*, Newark: University of Delaware Press, 1983.

Headrick, Daniel, *The Tools of Empire: Technology and European Imperialism in the Nineteenth Century*, Oxford: Oxford University Press, 1981.

Hichborn, H., *Report on European Dockyards*, Washington: Government Printing Office, 1886.

Higham, Robin, 'Government, Companies, and National Defence: British Aeronautical Experience, 1918–1945', *Business History Review* 39 (1965), 323–47.

Hill, Richard, *Egypt in the Sudan 1820–1881*, London: Oxford University Press, 1959.

Hirst, Francis W., *The Six Panics and Other Essays*, London: Methuen, 1913.

——, *The Political Economy of War*, London: Dent, 1916.

Hodges, Peter, *The Big Gun: Battleship Main Armaments, 1869–1945*, Greenwich: Conway Maritime Press, 1981.

Hogg, Ian V., *A History of Artillery*, London: Hamlyn, 1974.

——, *The Illustrated History of Ammunition from the Beginnings to the Present Day*, Syracuse, N.J.: Chartwell Books Inc., 1985.

Hogg, Ian and Batchelor, John, *Naval Gun*, Poole, Dorset: Blandford Press.

Hogg, O.F.G., *The Royal Arsenal: Its Background, Origin, and Subsequent History*, vol. II, London: Oxford University Press, 1963.

——, *Clubs to Cannon: Warfare and Weapons before the Introduction of Gunpowder*, London: Gerald Duckworth & Co., 1968.

——, *Artillery: Its Origin, Heyday and Decline*, London: C. Hunt, 1970.

Holmes, George, 'Naval Artillery', *The Army and Navy Magazine* 1 (1880), 206–13, 283–91.

Houghton, Walter (ed.), *The Wellesley Index to Victorian Periodicals 1825–1900*, 4 vols, Toronto: University of Toronto Press, pp. 1966–87.

Hovgaard, William, *Modern History of Warships*, London: E. and F.N. Spon, 1920; reprinted 1971.

Howarth, O.J.R., *The British Association For the Advancement of Science: A Retrospect 1831–1921*, London: B.A.A.S., 1922.

Hsu, Immanual C.Y., 'Late Ch'ing Foreign Relations, 1865–1905', in John K. Fairbank and Kwang-ching Liu (eds), *The Cambridge History of China*, vol. 11, *Late Ch'ing 1800–1911*, Part 2, Cambridge: Cambridge University Press, 1980, pp. 70–141.

——, *The Rise of Modern China*, 3rd edn, Oxford: Oxford University Press, 1983.

Hughes, Thomas, *Networks of Power: Electrification in Western Society, 1880–1930*, Baltimore: Johns Hopkins University Press, 1983.

——, 'Technological Momentum', in Merritt Roe Smith and Leo Marx (eds), *Does Technology Drive History? The Dilemma of Technological Determinism*, Cambridge: Mass.: MIT Press, 1987, pp. 101–13.

——, *American Genesis: A Century of Invention and Technological Enthusiasm, 1879–1970*, New York: Viking, 1989.

——, 'From Deterministic Dynamos to Seamless Web Systems', in H.E. Sladovich (ed.), *Engineering as a Social Enterprise*, National Academy Press, 1991, pp. 7–25.

Humble, Richard (ed.), *Naval Warfare: An Illustrated History*, Foreword by Peter Kemp, London: Orbis Publishing Limited, 1983.

Institution of Civil Engineers, *Sir W.G. Armstrong and Mr. J. Scott Russell: Correspondence and Documents Submitted to, with Minutes of Evidence Taken Before, The Council of the Institution of Civil Engineers*, London: Institution of Civil Engineers, 1867.

Irving, R.J., 'New Industries for Old? Some Investment Decisions of Sir W.G. Armstrong, Whitworth and Co. Ltd., 1900–1914', *Business History* 17 (July 1975), 150–69.

Jacques, W.H., *Modern Armor for National Defense*, New York: Putnam: 1886.

Jane, Fred T. (ed.), *Jane's All the World's Fighting Ships 1898*, London: Sampson Low Marston, 1898; reprinted Newton Abbot, Devon: David & Charles (Publishers) Limited, 1969.

Jenkins, Roy, *Asquith*, London: Collins, 1965.

Johnson, Curt, *The Big Guns Go To War*, Introduction by Aram Bakshian, Jr, London: Octopus Books Limited, 1975.

Johnson, Franklyn A., *Defence by Committee: The British Committee of Imperial Defence 1885–1959*, London: Oxford University Press, 1960.

Joll, James, *The Origins of the First World War*, London: Longman, 1984.

Kemp, Peter, 'From Tryon to Fisher: The Regeneration of a Navy', in Gerald Jordan (ed.), *Naval Warfare in the Twentieth Century, 1900–1945: Essays in Honour of Arthur Marder*, London: Croom Helm, 1977.

Kennedy, Paul, *The Rise and Fall of British Naval Mastery*, London: Allen Lane, 1976.

——, *The Rise of Anglo-German Antagonism, 1860–1914*, London: George Allen & Unwin, 1980.

——, *The Realities Behind Diplomacy: Background Influences on British External Policy, 1865–1980,* London: Fontana, 1981.

——, *Strategy and Diplomacy 1870–1945*, London: Fontana, 1983.

——, *The Rise and Fall of the Great Powers: Economic Change and Military Conflict from 1500 to 2000*, New York: Random House, 1987.

Kindleberger, Charles P., *Economic Growth in France and Britain 1851–1950*, Cambridge, Mass.: Harvard University Press, 1964.

King, James W., *The Warships and Navies of the World, 1880*, Introduction by N.J.M. Campbell, Annapolis, Md: Naval Institute Press, 1982; reprint of 1880 edn.

Kinglake, Alexander W., *The Invasion of the Crimea*, vol. 3, New York: Harper & Brothers, 1875: first published 1863.

Kinross, Lord, *Ataturk: The Rebirth of a Nation*, London: Weidenfeld and Nicolson, 1964.

Knight, C.N. (comp.), *War Office List*, London: Harrison and Sons, 1911.

Koch, H.W. *The Rise of Modern Warfare 1618–1815*, London: Bison Books, 1981.

Kondo, Adm. Motoki, 'The Progress of Naval Construction in Japan', *Transactions of the Institute of Naval Architects* 53 (1911, Part 2), 50.

Koss, Stephen, *The Rise and Fall of the Political Press in Britain*, vol. 1, *The Nineteenth Century*, London: Hamish Hamilton, 1981.

Kranzberg, Melvin, 'The Technical Elements in International Technological Transfer: Historical Perspectives', in John McIntyre and Daniel Papp (eds), *The Political Economy of International Technological Transfer*, New York: Quorum Books, 1986.

Krein, David F., *The Last Palmerston Government: Foreign Policy, Domestic Policy and the Genesis of `Splendid Isolation'*, Iowa: Iowa State University Press, 1978.

Kuo, Ting-yee, 'Self-strengthening: the Pursuit of Western Technology', in John K. Fairbank (ed.), *The Cambridge History of China*, vol. 10, *Late Ch'ing, 1900–1911*, Part 1, Cambridge: Cambridge University Press, 1978, pp. 491–542.

Lambert, Andrew, *Battleships in Transition: The Creation of the Steam Battlefleet 1815–1860*, Annapolis, Md: Naval Institute Press, 1984.

Landes, David, *The Unbound Prometheus: Technological Change and Industrial Development in Western Europe from 1750 to the Present*, Cambridge: Cambridge University Press, 1969.

Langer, William L., *The Diplomacy of Imperialism, 1890–1902*, 2nd edn, New York: Alfred A. Knopf, 1956.

Lavery, Brian, *The Ship of the Line*, vol. II; *Design, Construction and Fittings*, London: Conway Maritime Press, 1984.

LeFevour, Edward, *Western Enterprise in Late Ch'ing China: A Selective Survey of Jardine, Matheson and Company's Operations 1842–1895*, Cambridge, Mass.: Harvard University Press, 1968.

Liu, Kwang-ching, 'The Military Challenge: the North-West and the Coast', in John K. Fairbank and Liu Kwang-ching (eds), *The Cambridge History of China*, vol. II, *Late Ch'ing, 1800–1911*, Part 2, Cambridge: Cambridge University Press, 1980, pp. 202–73.

Lloyd, Christopher, *The Nation and the Navy*, London: Cresset Press, 1954.

Lloyd George, David, *War Memoirs*, vol I, *1914–1915*, Boston: Little, Brown, 1933.

Lyon, David, *The Ship: Steam, Steel and Torpedoes: The Warship in the 19th Century*, London: HMSO, 1980.

Lyon, Hugh, 'The Relations Between the Admiralty and Private Industry in the Development of Warships', in Brian Ranft (ed.), *Technical Change and British Naval Policy, 1860–1939*, London: Hodder and Stoughton, 1977, pp. 37–64.

Machlup, Fritz and Penrose, Edith, 'The Patent Controversy in the Nineteenth Century', *Journal of Economic History* X (May 1950), 1–29.

Mackay R.F., *Fisher of Kilverstone*, Oxford: Oxford University Press, 1973.

Mackenzie, John P., *The British Cabinet*, Toronto: University of Toronto Press, 1962.

MacLeod, Roy and Collins, Peter (eds), *The Parliament of Science: The British Association For the Advancement of Science 1831–1981*, London: Science Reviews, 1981.

Macrosty, Henry W., *The Trust Movement in British History: a study of business organization*, London: Longman, Green, 1907.

Magnus, Philip, *King Edward VIII*, Harmondsworth: Penguin, 1964.

Manchester, William, *The Arms of Krupp 1587–1968*, New York: Bantam, 1970.
Manning, Frederick, *The Life of Sir William White*, London: John Murray, 1923.
Marder, Arthur J., 'The English Armaments Industry and Navalism in the Nineties', *Pacific Historical Review* 7 (1938), 241–53.
——, *The Anatomy of British Sea Power: A History of British Naval Policy in the Pre-Dreadnought Era, 1880–1905*, New York: Alfred Knopf, 1940.
——, *From The Dreadnought to Scapa Flow: The Royal Navy in the Fisher Era, 1904–1919*, vol. 1, *The Road to War, 1904–1914*, London: Oxford University Press, 1961.
Martin, Kinsley, *The Triumph of Lord Palmerston: A Study of Public Opinion in England Before the Crimean War*, London: Hutchinson, 1924.
Matthews, H.C.G., *The Gladstone Diaries*, vol. 6, Oxford: Oxford University Press, 1978.
McGann, Thomas F., *Argentina, the United States, and the Inter-American System 1880–1914*, Cambridge, Mass.: Harvard University Press, 1957.
McHugh, Jeanne, *Alexander Holley and the Makers of Steel*, Baltimore: The Johns Hopkins University Press, 1980.
McKenzie, Peter, *W.G. Armstrong: The Life and Times of Sir William George Armstrong*, Morpeth: Longhirst, 1983.
McNeil, Ian, *Joseph Bramah: A Century of Invention, 1749–1851*, Newton Abbot: David & Charles, 1968.
McNeill, William H., *The Pursuit of Power: Technology, Armed Force, and Society Since A.D. 1000*, Chicago: University of Chicago Press, 1982.
Menne, Bernhard, *Krupp*, London: Wm. Hodge, 1937.
Miliband, Ralph, *The State in Capitalist Society: The Analysis of the Western System of Power*, London: Quartet, 1973.
Millman, Richard, *British Foreign Policy and the Coming of the Franco-Prussian War*, Oxford: Clarendon Press, 1965.
Mitchell, Donald W., *A History of Russian and Soviet Sea Power*, New York: Macmillan Publishing Co., Inc., 1974.
Moldelski, George, *Seapower in Global Politics 1494–1993*, London: Macmillan, 1988.
Morris, A.J.A., *The Scaremongers: the Advocacy of War and Rearmament 1896–1914*, London: Routledge & Kegan Paul, 1984.
Murray, H.R., *Krupp's and the International Armaments Ring: The Scandal of Modern Civilization*, London: Holden & Hardingham, 1916.

Needham, Joseph, with collaboration of Ho Ping-yu, Lu Gwei-Djen, *Science and Civilisation in China*, vol. 4, *Physics and Physical Technology*, Part 3, *Civil Engineering and Nautics*, Cambridge: Cambridge University Press, 1971.

Needham, Joseph, with collaboration of Wang Ling and Gwei-Djen, *Science and Civilisation in China*, vol. 5, *Chemistry and Chemical Technology*, Part 7, *Military Technology: The Gunpowder Epic*, Cambridge: Cambridge University Press, 1986.

Nelmes, Grahame V., 'Stuart Rendel and Welsh Liberal Political Organization in the Late-Nineteenth Century', *Welsh History Review* 9 (1978–9), 468–85.

Newbold, J.T.W., *The War Trust Exposed*, London: National Labour Press, 1913.

Newman, James R., *The Tools of War*, New York: Doubleday, Doran & Company, Inc., 1942.

Nicholas, Stephen J., 'The Overseas Marketing Performance of British Industry, 1870–1914', *Economic History Review* 37 (1984), 489–506.

Noble, Celia Brunel, *The Brunels: Father and Son*, London: Cobden-Sanderson, 1938.

Noble, David, *America By Design*, New York: Knopf, 1977.

——, *Forces of Production*, New York: Knopf, 1984.

Noel-Baker, Philip, *The Private Manufacture of Armaments*, London: Victor Gollancz, 1937.

Norman, E. Herbert, *Japan's Emergence as a Modern State: Political and Economic Problems of the Meiji Period*, New York: Institute of Pacific Relations, 1940.

Offner, Avner, 'Morality and Admiralty: "Jacky" Fisher, Economic Warfare and the Laws of War', *Journal of Modern History* 23 (1988).

Oman, Charles, *A History of the Art of War in the Sixteenth Century*, Elstree, Hertfordshire: Greenhill Books, 1987; reprint of 1937 edn.

Ono, Giichi, *Expenditures of the Sino-Japanese War*, New York: Oxford University Press, 1922.

——, *War and Armaments Expenditures of Japan*, New York: Oxford University Press, 1922.

Padfield, Peter, *The Battleship Era*, London: Pan, 1972.

——, *Guns at Sea*, London: Hugh Evelyn, 1973.

——, *The Great Naval Race: The Anglo-German Naval Rivalry, 1900–1914*, London: Hart-Davis, MacGibbon, 1974.

——, *Rule Britannia: The Victorian and Edwardian Navy*, London: Routledge & Kegan Paul, 1981.

Palin, H., *Memoir of the Late Lord Armstrong*, London: Gurney, 1901.

Parker, Foxhall A., *Squadron Tactics Under Steam*, New York: D. Van Nostrand, 1864.

Parker, Geoffrey, *The Military Revolution: Military Innovation and the Rise of the West, 1500–1800*, Cambridge: Cambridge University Press, 1988.

Parkes, Oscar, *British Battleships, 'Warrior' 1860 to 'Vanguard' 1950: A History of Design, Construction and Armament*, London: Seeley Service, 1966.

Parratt, Geoffrey, *The Royal Navy: The Sure Shield of the Empire*, London: The Sheldon Press, 1930.

Partridge, Michael S., 'Military Planning for the Defence of the United Kingdom 1814–1870', Ph.D thesis, London School of Economics, 1984.

Payne, Peter, 'The Emergence of the Large-scale Company in Great Britain, 1870–1914', *Economic History Review* XX (1967), 519–42.

——, *Colvilles and the Scottish Steel Industry*, Oxford: Clarendon Press, 1979.

Pearton, Maurice, *The Knowledgeable State: Diplomacy, War and Technology Since 1830*, London: Burnett, 1982.

Peebles, Hugh, *Warshipbuilding on the Clyde: naval orders and the prosperity of the Clyde shipbuilding industry, 1889–1939*, Edinburgh: J. Donald, 1987.

Pellow, Pownall, *On Fleet Maneuvering*, London: 1868.

Penrose, Edith T., *The Theory of the Growth of the Firm*, London: Basil Blackwell, 1959.

Perris, George H., *The War Traders: An Exposure*, London: National Peace Council, 1913.

Petrovic, Djurdjica, 'Fire-arms in the Balkans on the Eve of and After the Ottoman Conquests of the Fourteenth and Fifteenth Centuries', in V.J. Parry and M.E. Yapp (eds), *War, Technology and Society in the Middle East*, London: Oxford University Press, 1975.

Platt, D.C.M., *Finance, Trade and Politics in British Foreign Policy 1815–1914*, Oxford: Clarendon Press, 1968.

Pollard, Sidney, 'Laissez-Faire and Shipbuilding', *Economic History Review* 5 (1952), 98–115.

Pollard, Sidney and Robertson, Paul, *The British Shipbuilding Industry: 1870–1914*, Cambridge, Mass.: Harvard University Press, 1979.

Preston, Antony, *Battleships*, New York: Cresset Books, 1981.

——, *History of the Royal Navy*, London: Hamlyn, 1983.

Purves, J.G., 'Nineteenth-Century Russia and the Revolution in Military Technology,' J.G. Purves and D.A. West (eds), *War and Society in the Nineteenth-Century Russian Empire*, Toronto: New Review Books, 1972.

Ranft, Bryan, 'The Protection of British Seaborne Trade and the Development of Systematic Planning for War, 1860–1906', in Bryan

Ranft (ed.), *Technical Change and British Naval Policy, 1860–1939*, London: Hodder and Stoughton, 1977.

Rawlinson, John L., *China's Struggle for Naval Development, 1839–1895*, Cambridge, Mass.: Harvard University Press, 1967.

Reader, W.J., *Imperial Chemical Industries: A History*, London: Oxford University Press, 1970.

Rendel, Stuart, Lord, *The Personal Papers of Lord Rendel*, ed. F.E. Hamer, London: Ernest Benn, 1931.

Ripley, Warren, *Artillery and Ammunition of the Civil War*, New York: Van Nostrand Reinhold Company, 1970.

Rippon, P.M., *Evolution of Engineering in the Royal Navy*, vol. I, *1827–1939*, Tunbridge Wells, Kent: Spellmount Limited, 1988.

Ritter, Gerhard, *The Sword and the Sceptre: The Problem of Militarism in Germany*, vol. II, *The European Powers and the Wilhelminian Empire, 1890–1914*, trans. Heinz Norden, Coral Gables, Florida: University of Miami Press, 1965.

Robertson, Frederick, *The Evolution of Naval Armament*, London: Constable, 1921.

Rolt, L.T.C., *Isambard Kingdom Brunel: A Biography*, London: Green, 1957.

——, *Victorian Engineering*, Harmondsworth, Middlesex: Pelican Books, 1974.

Ropp, Theodore, *War in the Modern World*, rev. edn, New York: Collier Books, 1962; first published 1959.

——, *The Development of a Modern Navy: French Naval Policy, 1871–1904*, Annapolis, Md.: Naval Institute Press, 1987; written in 1937.

Roskill, S.W., *The Strategy of Sea Power: Its Development and Application*, London: Collins, 1962.

Rothenberg, Gunther E., *The Art of War in the Age of Napoleon*, London: B.T. Batsford, 1977.

Salevouris, Michael J., *Riflemen Form: The War Scare of 1859–1860 in England*, New York: Garland, 1982.

Sampson, Anthony, *The Arms Bazaar: The Companies, The Deals, The Bribes: From Vickers to Lockheed*, London: Coronet, 1977.

Sanderson, Michael, 'The Professor as Industrial Consultant: Oliver Arnold and the British Steel Industry, 1900–14', *Economic History Review* 31 (1978), 585–600.

Sandler, Stanley, *The Emergence of the Modern Capital Ship*, Newark, Delaware: University of Delaware Press, 1979.

Schumpeter, Joseph, *The Theory of Economic Development*, Cambridge, Mass.: Harvard University Press, 1934.

Schurman, D.M., *The Education of a Navy: The Development of British Naval Strategic Thought, 1867–1914*, Malbar, Florida: Robert E. Keiger, 1984.

Scott, J.D., *Vickers: A History*, London: Weidenfeld & Nicolson, 1962.

Segreto, Luciano, 'More Trouble Than Profit: Vickers' Investments in Italy 1906–39', *Business History* 27 (1985), 316–37.

Seldes, George, *Iron, Blood and Profits: An Exposure of the World-Wide Munitions Racket*, New York: Harper, 1934.

Semmel, Bernard, *Liberalism & Naval Strategy: Ideology, Interest, and Sea Power During the Pax Britannica*, Winchester, Mass.: Allen & Unwin, Inc., 1986.

Seton-Watson, R.W., *A History of the Roumanians*, London: Archon, 1963; first published 1934.

Sheppard, E. (ed.), *George Duke of Cambridge: A Memoir*, vol. 1, London: Longman, Green, 1906.

Singer, Charles, *et al., A History of Technology*, 8 vols, Oxford: Clarendon Press, 1954–84.

Smith, D. Bonner and Lumby, E.W.R., *The Second China War 1856–1860*, London: Naval Records Society, 1954.

Smith, Merritt Roe and Marx, Leo (eds), *Does Technology Drive History? The Dilemma of Technological Determinism*, Cambridge, Mass.: MIT Press, 1987.

Smith, Steven R.B., 'Public Opinion, the Navy and the City of London: The Drive for British Naval Expansion in the Late Nineteenth Century', *War and Society* 9:1 (May 1991), 29–50.

Soley, James Russell, *Report on Foreign Systems of Naval Education*, Washington: Government Printing Office, 1880.

Sprout, Margaret T., 'Mahan: Evangelist of Sea Power', in Edward Mead Earle (ed.), *Makers of Modern Strategy: Military Thought from Machiavelli to Hitler*, Princeton: Princeton University Press, 1971, pp. 415–45; first published 1943.
[Note the text listed above edited by Peter Payne is not identified as a revised or 2nd edition of the 1943 text, but as a separate book – note different sub-titles.]

Stanmore, Arthur Hamilton-Gordon, Lord, *Sidney Herbert of Lea: A Memoir*, 2 vols, London: Murray, 1906.

Staudenmaier, John, *Technology's Storytellers: Reweaving the Human Fabric*, Cambridge, Mass.: MIT Press, 1985.

Steinberg, Jonathan, *Yesterday's Deterrent: Tirpitz and the Birth of the German Battle Fleet*, New York: Macmillan, 1965.

Stoney, F.S. and Jones, C., *A Text-Book of the Construction and Manufacture of the Rifled Ordnance in the British Service*, 2nd edn, London: H.M.S.O., 1872.

Strachan, Hew, *European Armies and the Conduct of War*, London: George Allen & Unwin, 1983.

Sumida, Jon Tetsuro, *In Defence of Naval Supremacy: Finance, Technology, and British Naval Policy, 1889–1914*, London: Routledge, 1993.

Swartz, Marvin, *The Politics of British Foreign Policy in the Era of Disraeli and Gladstone*, London: Macmillan, 1985.

Tennent, James Emerson, *The Story of the Guns*, Richmond, Surrey, England: The Richmond Publishing Co., 1972; first published in 1864.

Thompson, Neville, *Wellington After Waterloo*, London: Routledge & Kegan Paul, 1986.

Trebilcock, Clive, 'A "Special Relationship" – Government and the Cordite Firms', *Economic History Review* 19 (1966), 364–79.

——, '"Spin-Off" in British Economic History: Armaments and Industry, 1760–1914', *Economic History Review* 22 (1969), 474–90.

——, 'Legends of the British Armament Industry 1890–1914: A Revision', *Journal of Contemporary History* 5 (1970), 3–19.

——, 'British Armaments and European Industrialization, 1890–1914', *Economic History Review* 26 (1973), 254–72.

——, 'War and the Failure of Industrial Mobilization: 1899 and 1914', in J.M. Winter (ed.), *War and Economic Development*, London: 1975.

——, 'The British Armaments Industry 1890–1914: False Legend or True Utility', in Geoffrey Best and Andrew Wheatcroft (eds), *War, Economy and the Military Mind*, London: Croom Helm, 1976, pp. 89–107.

——, *The Vickers Brothers: Armaments and Enterprise 1854–1914*, London: Europa, 1977.

Trotter, Wilfrid Pym, *The Royal Navy in Old Photographs*, London: J.M. Dent & Sons Ltd, 1975.

Ushisaburo, Kobayashi, *Military Industries of Japan*, New York: Oxford University Press, 1922.

Very, Edward W., *The Development of Armor for Naval Use*, Annapolis, Md.: Naval Institute Press, 1883.

Walpole, Spencer, *The History of Twenty-Five Years*, London: Longman, Green, 1904.

Warren, Kenneth, *Armstrongs of Elswick: Growth in Engineering and Armaments to the Merger with Vickers*, London: Macmillan, 1989.

Webster, Richard A., *Industrial Imperialism in Italy 1908–1915*, Berkeley: University of California Press, 1975.

Wedgwood, Josiah C., *Memoirs of A Fighting Life*, London: Hutchinson, 1941.

Weinroth, H.S., 'Left-wing Opposition to Naval Armaments in Britain Before 1914', *Journal of Contemporary History* 6 (1971), 93–121.

White, Sir William, 'History of the Institution of Naval Architects and of Scientific Education in Naval Architecture', *Transactions of the Institute of Naval Architects* 53 (2) (July 1911).

Wilson, H.W., 'The Protection of Our Commerce in War', *Nineteenth Century*, February 1896.

Wilson, Herbert Wrigley, *Ironclads in Action: a sketch of naval warfare from 1855 to 1895*, London: S. Low, Marston and Company, 1896.

Wilson, John, *C.B.: A Life of Sir Henry Campbell-Bannerman*, London: Constable, 1973.

Wilson, John F., *British Business History, 1720–1994*, New York: Manchester University Press, 1995.

Woodward, E.L., *Great Britain and the German Navy*, London: Frank Cass and Co. Ltd, 1964; first published 1935.

Wright, Stanley F., *Hart and the Chinese Customs*, Belfast: William Mullan & Son, 1950.

Wrottesley, George, *Life and Letters of Sir John Burgoyne*, London: Richard Bentley & Son, 1873.

Yamagata, Aritomo, 'The Japanese Army', in Shigenobu Okuma, *Fifty Years of New Japan*, vol. 1, ed. Marcus B. Huish, London: Smith, Elder, & Co., 1910, pp. 194–217.

Yamamoto, Gombey, 'The Japanese Navy', in Shigenobu Okuma, *Fifty Years of New Japan*, vol. 1, ed. Marcus B. Huish, London: Smith, Elder, & Co., 1910, pp. 218–230.

Yexley, Lionel, *The Inner Life of the Navy*, London: Sir Isaac Pitman & Sons, 1908.

Index

The author acknowledges and thanks Peter Walker for contributing to this index.